Environmental Microbiology

Environmental Microbiology

Annette Bolger

Oxford Book Company
Jaipur ▪ New Delhi

Published By

Oxford Book Company

Regd. Off. "Bony Residency", Gate No. 2
Opp. Tilak Public School, Vishweswariya Nagar,
Gopalpura Road, Jaipur - 302018, India
Ph: 0141 - 2761280, Telefax: 2761381
email - oxfordbookjpr@gmail.com
website - oxfordbookcompany.com

Delhi Office: 102, 1st Floor, Satyam House,
4327/3, Ansari Road, Darya Ganj,
New Delhi - 110002, India
Ph: 011-45652440

Distributed By

KEYA INTERNATIONAL PRESS LLP
E-mail: keyainternationalpressllp@gmail.com

Published in India, 2024

Rs. 1550

ISBN : 978-81-80179-71-1

Typeset by Shivangi Computers, Jaipur

Printed and bound in India by Thomson Press Ltd.,
New Delhi

Preface

Environmental Microbiology, besides a traditional discipline in Developing fast, because of realization of its importance in Industry, Agriculture Pharmaceutical concerns, Public Health, Geological explorations, bioenergetics and as a mean to exploit new sources of energy useful for various purposes. Consequently, nations are striving hard to merge the biotechnological operation into national development, seeking strategies for International cooperation and ties.

Environmental pollution control microbiology is an exciting and challenging area of science and engineering. It is concerned with protecting people from pathogenic micro-organisms and the application of microbiology to solve a wide range of environmental pollution problems. Both science and engineering are involved in pollution control. Scientists determine the environmental problems and evaluate different solutions, whereas engineers utilize the information supplied by scientists to design the most efficient systems to solve the environmental problems.

The concept of environmental pollution control microbiology has its roots in conventional microbiology with concerns for pathogenic micro-organisms and public health. It can even be traced in chemistry, which forms the basis of all reactions between chemical compounds that make up our world, and in biochemistry, which focuses on the chemistry of biological systems. In fact, microbiology encompasses several scientific fields such as bacteriology, virology, mycology, phycology, protozoology, zoology, biotechnology and mathematics.

Annette Bolger

Contents

Contents

Chapter 1

Biology of Bacterial Bioluminescence

LUMINOUS BACTERIA - HABITATION AND TAXONOMY

Luminous bacteria are the most widely distributed light-emitting organisms with the majority existing in seawater and the remainder living in the terrestrial or freshwater environment.

While most species of luminescent bacteria are capable of living free, the majority are found in nature associated in symbiosis with host organisms.

In symbiosis the bacteria are nourished with readily available food sources for growth, and at the same time the host utilizes the adopted illumination to communicate, to attract prey, and to masquerade itself from predators.

However, there are certain species of luminescent bacteria, which are obligatory symbionts, requiring unique nutritional supplements, which are exclusively available from the host. Although the presence of these obligatory symbionts have been detected, they are not separable from their host and therefore are unable to be cultured in the laboratory for further study. There are three major genera, into which most luminous bacteria are classified; *Photobacterium, Vibrio,* and *Photorhabdus*.

Species existing in the marine environment are mainly categorized into the *Photobacterium* and *Vibrio* genera, and the terrestrial species are classified into the *Photorhabdus* (previously designated as *Xenorhabdus*) genus. Species within the *Photobacterium* genus are generally light organ symbionts of marine animals, whereas the *Vibrio* species exist as free-living forms as well as symbionts in the sea.

Utilization of the Illumination Function is Believed to Frighten Nearby Predators Allowing the Squid to escape.

Many luminous bacteria are parasitic with *Photobacterium* and *Vibrio* families infecting marine crustacea, and *Photorhabdus luminescens* infecting terrestrial insects, such as caterpillars with nematodes as the intermediate host for the bacteria. In addition, free-living luminous bacteria that are dispersed in the seawater can often be found in both the gut tract and on the skin surface of almost all marine animals as non-specific parasites.

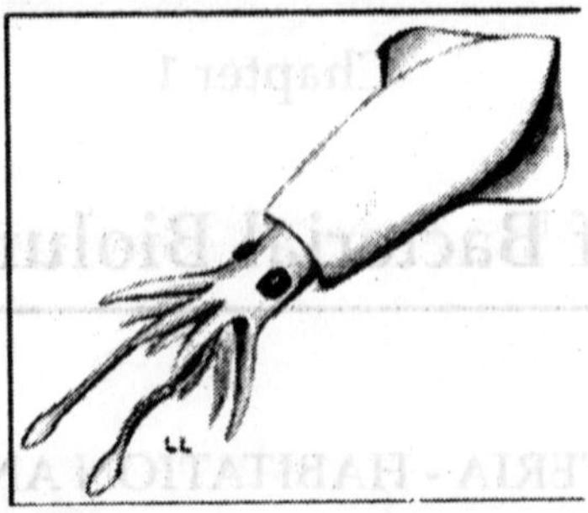

Fig. Luminous Bacteria Reside in Symbiosis on a Pair of Light Organs in the Mantle Body of the Squid.

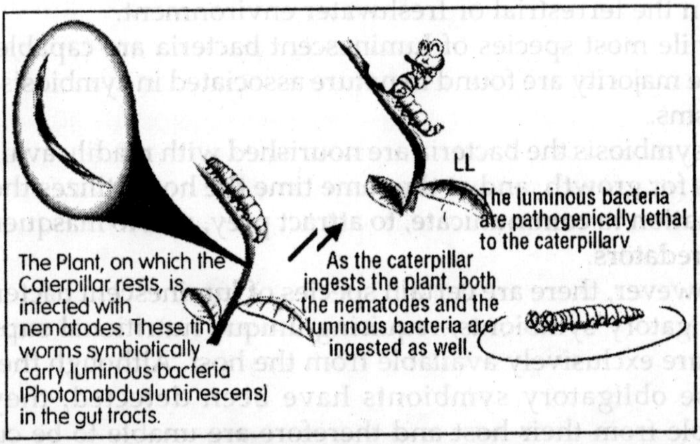

Fig. Infection with the Terrestrial Luminous Bacteria (*Photohabdus Luminescens*) is Pathogenically Lethal to the Caterpillar.

For luminous bacteria residing within the gut tracts of marine animals, the extra-cellular chitinase produced on the cell wall of all luminous bacteria facilitate the decomposition of the ingested chitin (e.g. from the exoskeletons of crustacea).

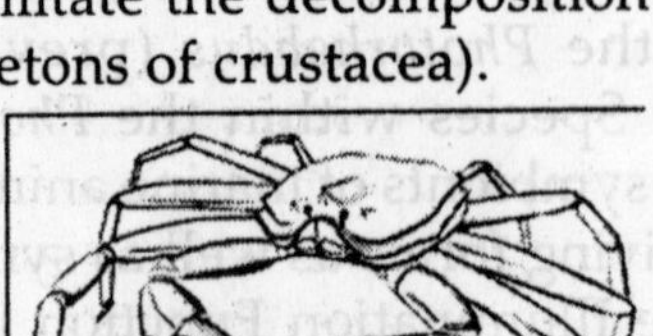

Fig. Luminous Bacteria are Non-Specific Parasites of Crustacea

The exoskeletons of Tanner crabs are the sites of colonization of luminous bacteria, whose infection causes lesions on the surface of its appendages.

Each species of luminous bacteria differs in a number of properties, including the specific growing conditions (nutritional requirements and growth temperature), and the reaction kinetics of the luciferase involved in light generation; however, all luminous bacteria are rod-shaped, gram-negative microorganisms with flagella facilitating motion.

Luminous bacteria are also facultative anaerobes capable of growth when the supply of molecular oxygen is limited. Despite the physiological diversity among different species of luminous bacteria, all luminescent microorganisms utilize highly homologous biochemical machineries to produce light. The onset as well as the energy output of this light-producing molecular machinery are tightly regulated under a central signaling pathway.

The essential biochemical components of bacterial bioluminescence, and the sequential molecular interactions/reactions along the reaction pathway. The final sections will walk briefly through the signaling pathways, by which the induction and the repression of the illumination function in luminescent bacteria are carefully controlled by cellular factors and intercellular communication.

Biochemistry of the Bacterial Bioluminescence Reaction – Bacterial Luciferase and the Light Colour

Bacterial luciferase is the enzyme that catalyzes light emission at the heart of bacterial bioluminescence. However, the catalytic machinery involved in continuous light production in luminous bacteria includes not only bacterial luciferase but also the enzymes that supply and regenerate the substrates of bacterial luciferase.

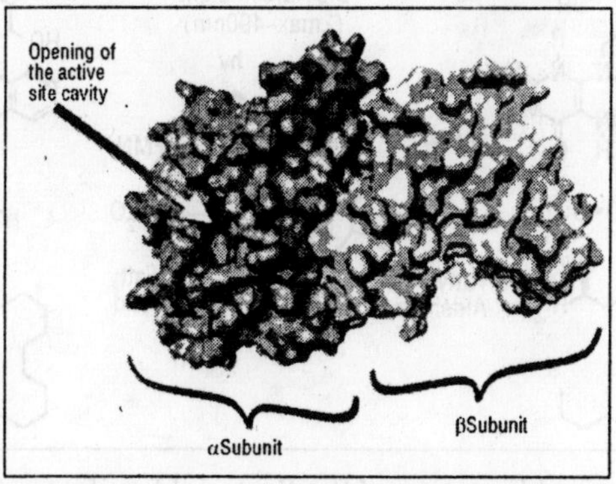

Fig. Bacterial Luciferase Structure.

The DNA sequences coding the proteins in the luminescent system are termed the *lux* genes. Bacterial luciferase is a heterodimer, composed of two different polypeptides, designated α and ß (of molecular mass 40 kDa and 37 kDa, respectively) and encoded by the *lux*A and *lux*B genes, respectively. The active site is located within the subunit.

In the absence of the β subunit, the α subunit alone functions inefficiently with a poor light yield.

As the crystal structure of *V. harveyi* luciferase reveals extensive interactions and complex binding patterns between several side chains and backbone amides of the and ß subunits, the function of the ß subunit has been assumed to act as a supporting scaffold assisting in the conformational change of the subunit during the catalysis.

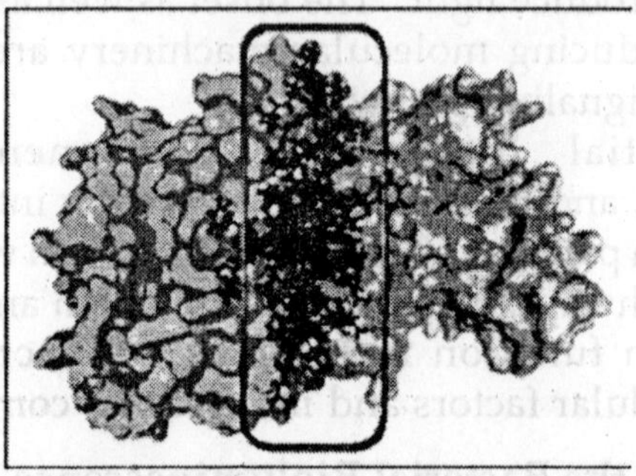

Fig. The Network of Extensive Inter-Subunit Interactions.

The substrates of bacterial luciferase are reduced flavin mononucleotide ($FMNH_2$), molecular oxygen, and long chain fatty aldehyde.

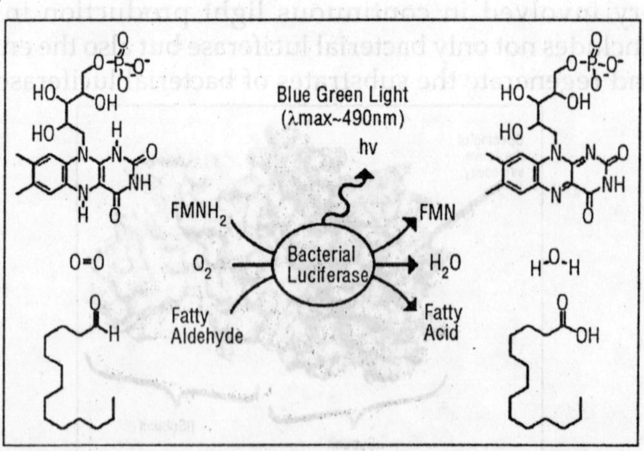

Fig. The Net Chemical Equation of the Bacterial Luciferase Catalyzed Reaction.

The excess energy, which is liberated from the oxidation of $FMNH_2$

and aldehyde concomitant with the reduction of molecular oxygen, is released as blue/green light emission ($_{MAX}$ ~ 490 nm).

Luciferases function well with long chain aldehydes of eight carbons or longer with tetradecanal presumed to be the natural aldehyde used *in vivo* in the luminescent bacteria.

The characteristic colour indicates the energy level of the photon that was produced when the excited electron on the flavin chromophore returns to the ground state. Researchers in the field have discovered that flavin analogs with substituted atoms in the chromophore moiety resulted in different luciferase emission colors.

In addition, it has also been shown that point mutations at the flavin chromophore's binding site distorts the colour emission spectrum of bacterial bioluminescence, indicating that the distinctive emission colour depends not only on the chromophore that emits the photon, but also the electronic nature of the chromophore-binding microenvironment in luciferase.

Aside from bacterial luciferase, some luminous bacteria carry fluorescent proteins to modulate the emission colour, distinguishing themselves from other strains. Interaction of blue fluorescent protein (also known as the lumazine protein) in *Photobacterium phosphoreum* and *Photobacterium leiognathi* with the respective luciferases yields photon emission at higher energy ($_{max}$ ~ 478nm) corresponding to a blue colour.

Similarly, the presence of the yellow fluorescence protein in the strain Y–1 of *V. fischeri* resulted in yellow light emission ($_{max}$ ~ 545nm). The participation of the fluorescence proteins in the luciferase-catalyzed reaction alters the reaction kinetics of bacterial luciferase yielding a much higher quantum efficiency than can be achieved by luciferase alone.

Although the molecular mechanism is not fully understood, researchers are in favour of an energy transfer model, in which the fluorescent proteins receive the energy from a high-energy luciferase reaction intermediate rather than absorb the blue-green light emitted at 490 nm.

While different colors of light emission can be achieved through energy transfer to the different fluorescent proteins, the energy input is clearly provided by the oxidation of $FMNH_2$ and aldehyde by bacterial luciferase. In order for light emission to occur for prolonged period of times, substrates must be supplied continuously to bacterial luciferase.

As the time frame and the level of product generation in enzymatic reactions are limited by the availability of the substrates, the constant light emission in luminous bacteria must therefore be maintained by several different enzymes continuously generating the substrates for the bioluminescence reaction.

Those enzymes that replenish the aldehyde substrate are coded on the *lux* operon; in particular, the fatty acid reductase, a multienzyme complex, whose *lux* genes (*lux*C, *lux*D, and *lux*E) immediately flank the *lux*A and *lux*B genes of luciferase. Other genes including *lux*F, *lux*G, and *lux*H whose functions are neither clearly defined nor apparently necessary for bioluminescence are also found in some *lux* operons.

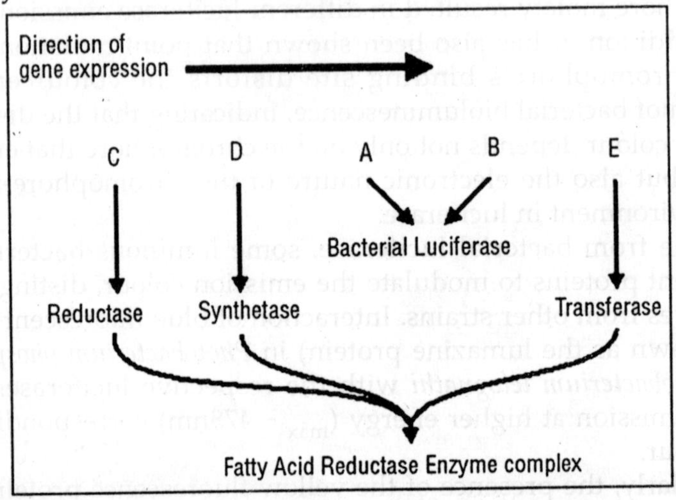

Fig. The Arrangement of the *Lux* CDABE Open Reading Frames.

FATTY ALDEHYDE SYNTHESIS AND RECYCLING IN LUMINOUS BACTERIA – FATTY ACID REDUCTASE

Although a structural model of the fatty acid reductase at atomic resolution to visualize how fatty acid is exchanged among subunits of this multienzyme complex and modified toward the fatty aldehyde in three-dimensional space is not available, separation and isolation of fatty acid reductase into individual subunits has allowed the identification of the function of each component in the protein complex. Fatty acid reductase is a large protein with approximately 500 kilodaltons of molecular mass.

This large complex is comprised of 12 polypeptides (four polypeptides coded by each of the *lux*C, *lux*D, and *lux*E genes). Although the designated letters for this set of *lux* genes go from C to E, the progression of aldehyde synthesis does not follow the same direction. *Lux*D encodes the transferase, which transfers the acyl moiety of fatty acyl-ACP, acyl-CoA, and other activated acyl donors derived from the universal fatty acid biosynthetic pathway, to the hydroxyl group of a serine on the transferase, followed by the conversion of the ester to a fatty acid through hydrolysis.

The transferase has a high specificity for acyl derivatives with chain lengths of fourteen carbons, consistent with tetradecanal being the aldehyde used in the bioluminescent reaction "in vivo."

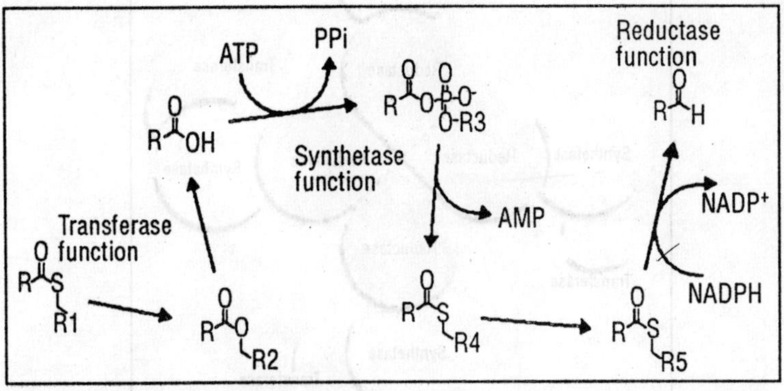

Fig. The Channeling of Substrates Among the Transferase (blue), the Synthetase (yellow), and the Reductase (green) in the Fatty Acid Reductase Enzyme Complex.

The fatty acid produced by the transferase is then channeled to the synthetase (LuxE), which activates the carboxylate function with ATP to form a mixed anhydride of acyl-AMP. The acyl group on the anhydride intermediate, acyl-AMP, then reacts with the reactive thiol of cysteine in the synthetase, forming a covalent linkage between the acyl group and the synthetase and release of AMP.

Through a reciprocal trans-esterification reaction between a thiol group on the synthetase and a thiol group on the reductase (LuxC), the fatty acyl group exchanges the covalent bonding partner from synthetase to reductase.

The reductase then breaks the thiol-ester linkage by reduction with the hydride of NADPH to release fatty aldehyde and return itself to its original free enzyme state. The subunits of fatty acid reductase are organized in a radial fashion with a central core of four reductase subunits, which are weakly complexed with each individual synthetase subunit.

The transferases (LuxD) branch from the synthetase at the outermost periphery in the fatty acid reductase complex, and dissociate very readily.

The advantage of such complex formation is the efficient channeling of the reaction intermediate from within one subunit to the other without the exposure of the hydrophobic hydrocarbons of the reaction intermediates into the aqueous cytosolic environment of the luminous bacteria.

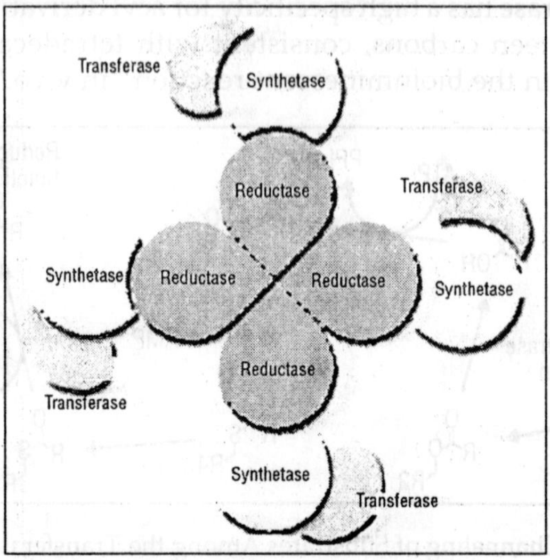

Fig. The Association Map of the Transferase, Synthetase, and Reductase Units in the Fatty Acid Reductase Enzyme Complex.

In addition, the arrangement of transferase, synthetase, and reductase in a radial fashion channels the reaction material through different microenvironments that stabilize the acyl intermediate at various stages of reaction as the material is processed from the aqueous periphery, where the acyl group is received, towards the centre of the complex where the reductase reduces the fatty acyl group to aldehyde.

As all proteins in the fatty acid reductase complex have a high specificity for tetradecanoyl groups, the final product is tetradecanal.

RECHARGING THE $FMNH_2$ – FMN REDUCTASE

Flavin mononucleotide (FMN) is an electron carrier present in the electron transport chains of both eukaryotic and prokaryotic cells. FMN is capable of donating and receiving one electron through radical formation or two electrons with hydride transfer.

FMN, the oxidized state, is highlighted within the yellow oval. FMNH•, the half reduced or the radical state, is highlighted in the blue oval. $FMNH_2$, the fully reduced or the hydroquinone state, is highlighted in the pale oval. FMN is derived from riboflavin (also known as Vitamin B_2), and both riboflavin and FMN are common nutrients essential for the survival of both eukaryotic and prokaryotic cells.

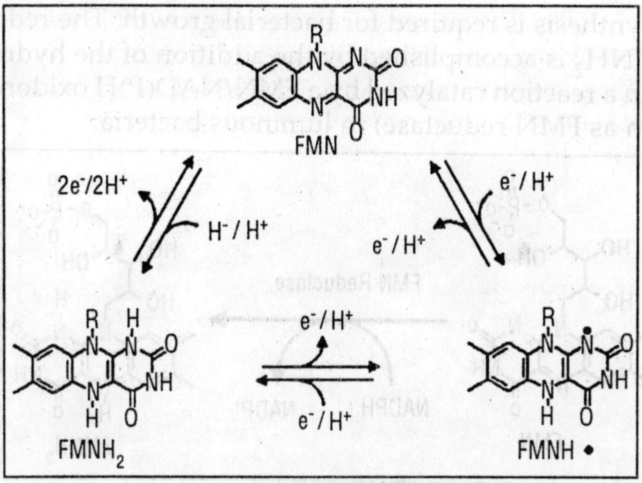

Fig. The 3 Oxidation States of Flavin Mononucleotide

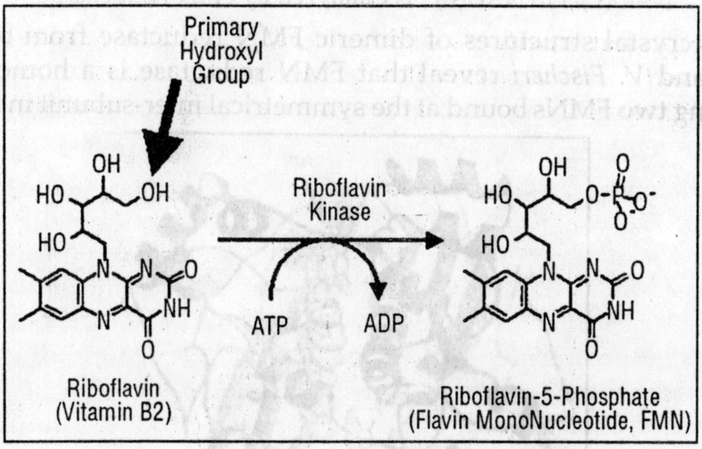

Fig. The Transfer of the Phosphate Group from ATP to Riboflavin.

The blue arrow highlights the primary hydroxyl group, to which riboflavin kinase adds the phosphate, and distinguishes it from the other secondary hydroxyl groups on the ribityl side chain.

The fact that the conversion of a nonluminous bacterium, such as *Escherichia coli,* to a light-emitter requires only the insertion of the *lux*CDABE genes, encoding the bacterial luciferase and the fatty acid reductase complex, into the cell, indicates that $FMNH_2$ is readily provided from the electron transport chain in all bacteria. Although the biosynthesis of riboflavin and FMN in luminous bacteria is carried out in multiple steps by enzymes that are not encoded by the *lux* gene system, these enzymes are generally present in bacteria, as riboflavin and FMN synthesis is required for bacterial growth. The reduction of FMN to $FMNH_2$ is accomplished by the addition of the hydride from NAD(P)H in a reaction catalyzed by a FMN/NAD(P)H oxidoreductase (also known as FMN reductase) in luminous bacteria.

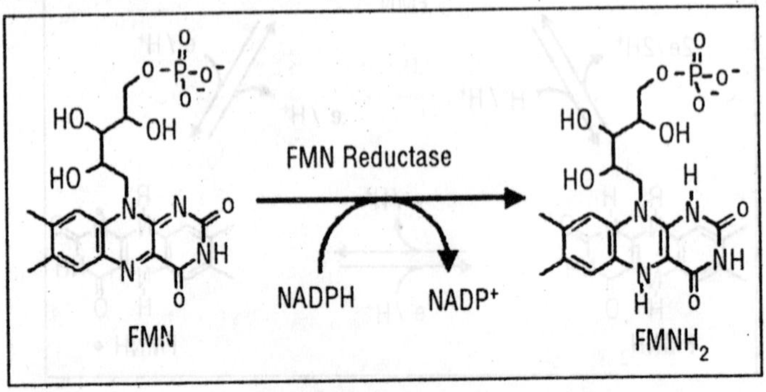

Fig. The Reduction of FMN to $FMNH_2$, Coupled with the Oxidation of NAD(P)H to $NAD(P)^+$, is Catalyzed by FMN Reductase.

The crystal structures of dimeric FMN reductase from both *V. harveyi* and *V. Fischeri* reveal that FMN reductase is a homodimer containing two FMNs bound at the symmetrical inter-subunit interface.

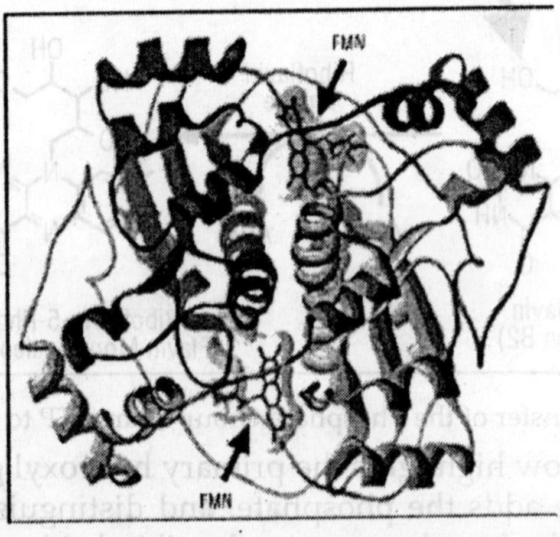

Fig. Two FMNs, which are Highlighted in Yellow Clouds, are Embraced at the Inter-Subunit Interface Between the Two Identical Subunits that are Colored in Red and Green.

Recently, researchers in the field have carried out extensive kinetic investigations in an attempt to elucidate how $FMNH_2$ is channeled from the FMN reductase to the bacterial luciferase. Based on the strong binding affinities of FMN reductase toward both FMN and $FMNH_2$, and the tightly anchored FMN binding geometries in the crystal structure of FMN reductase, investigators have speculated that the transfer of $FMNH_2$ from FMN reductase to bacterial luciferase occurs by transient formation of a luciferase-FMN reductase complex in luminous bacteria rather than through free diffusion of $FMNH_2$ dispatched into the cytosol.

This would result in a very efficient recycling and reduction of FMN to $FMNH_2$ for use by bacterial luciferase.

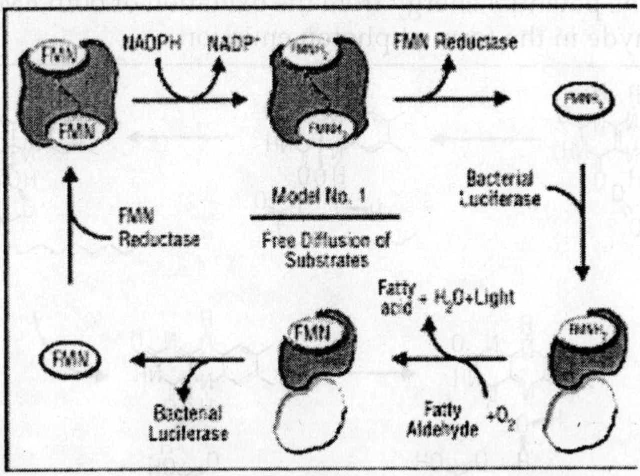

Fig. The Uptake of Substrate from the Cytosolic Environment is One of the Two Possible Modes, by Which Bacterial Luciferases have Access to $FMNH_2$.

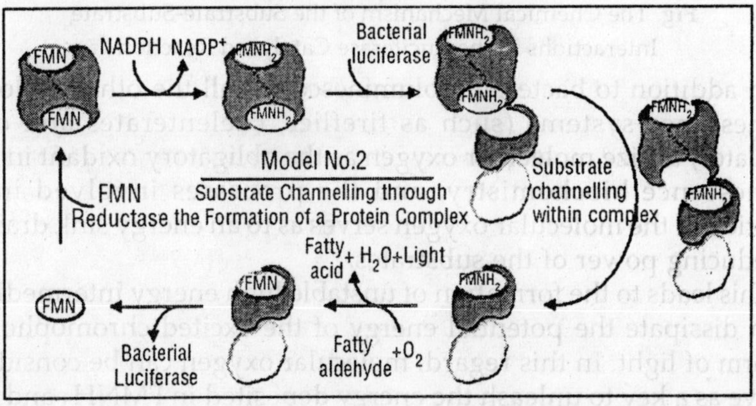

Fig. The Direct Channeling of the $FMNH_2$ from FMN Reductase to the Bacterial Luciferase within the Transient Reductase-Luciferase Complex is the Second Possible Mode of Substrate Delivery to the Luciferase.

THE UNIVERSAL OXIDANT IN BIOLUMINESCENCE – MOLECULAR OXYGEN

A critical component in the biochemistry of bacterial

bioluminescence is molecular oxygen, which is supplied from the external cellular environment. Without the input of molecular oxygen, luminous bacteria cannot emit light.

In the bacterial luciferase-catalyzed reaction, the energy expenditure on the reduction of molecular oxygen to a peroxy reaction intermediate and then ultimately to water serves as a trigger for releasing the potential energy from the oxidation of both $FMNH_2$ and fatty aldehyde in the form of photon emission.

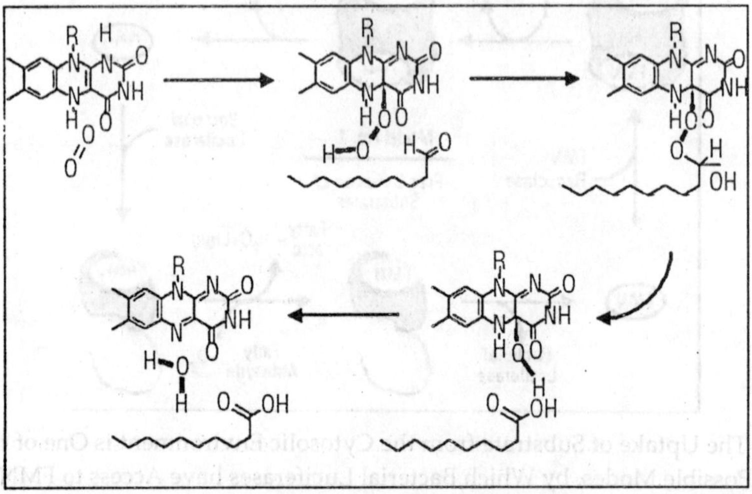

Fig. The Chemical Mechanism of the Substrate-Substrate Interactions in the Luciferase Catalyzed Reaction.

In addition to bacterial bioluminescence, all the other biological luminescence systems (such as fireflies, coelenterates and dino-flagellates) utilize molecular oxygen as the obligatory oxidant in their luminescence biochemistry, and the processes involved in the reduction of the molecular oxygen serves as to an energy sink draining the reducing power of the substrates.

This leads to the formation of unstable high energy intermediates, which dissipate the potential energy of the excited chromophore in the form of light. In this regard, molecular oxygen can be considered to serve as a key to unleash the energy deposited in $FMNH_2$ and fatty aldehyde for bacterial bioluminescence.

THE UTILIZATION OF BACTERIAL BIOLUMINESCENCE AS A BIOSENSOR AND REPORTER OF GENE EXPRESSION

As the primary function of bacterial luciferase is to catalyze the emission of light, this feature together with generation of the aldehyde substrate by fatty acid reductase can be successfully produced in other bacteria, by transfer of the *lux*CDABE genes and converting nonluminescent bacteria into light emitters.

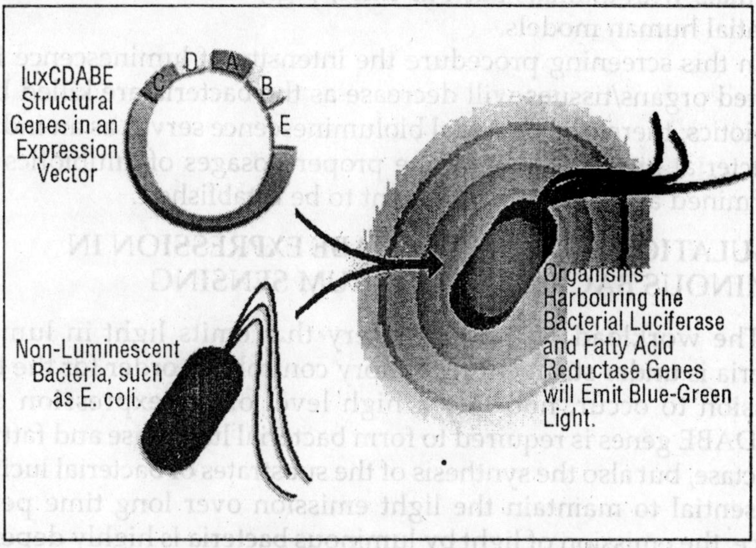

Fig. The "insertion" of the Foreign *lux*CDABE Structural Genes into the Organism Allows the Fatty Acid Reductase Enzyme Complex and Luciferase to be Expressed, and the Function of Luciferase Confers the Organism the Ability to Emit Light.

The adaptation of light as an accessory feature of normally nonluminescent bacteria has provided researchers an easy alternative to measure and detect the growth and living conditions of bacteria. The utilization of bacterial bioluminescence is also one of several methods used in the detection of pathogenic bacteria in human food sources.

By culturing a food sample in the presence of a recombinant bacteriophage carrying the *lux*CDABE insert, one can readily determine the presence of bacterial contamination in the food source. In addition, the light emitting property of the *lux*CDABE genes has been employed as a reporter of gene expression for studying regulatory controls involved in affecting the efficiency of the RNA polymerase in initiation and transcription at different promoters.

When the *lux*CDABE genes are under the control of an environmentally regulated promoter (e.g. promoters whose efficiency is highly sensitive to the level of mercury, arsenic, or other pollutants), the structural *lux* genes can function as a biosensor, whose expression will monitor the presence of toxic waste in the environment.

In the pharmaceutical industry, genetically modified bacteria carrying the *lux* luminous system have been utilized to evaluate the efficacy of antibiotics in fighting against bacterial infections in mammals with animals such as mice, pigs, and monkeys serving as potential human models.

In this screening procedure the intensity of luminescence in the infected organs/tissues will decrease as the bacteria are killed by the antibiotics; therefore, bacterial bioluminescence serves as an indicator of bacterial growth allowing the proper dosages of antibiotics to be determined and effective treatment to be established.

REGULATION OF THE LUXCDABE EXPRESSION IN LUMINOUS BACTERIA – QUORUM SENSING

The workload of the machinery that emits light in luminous bacteria is under stringent regulatory controls. In order for the bright emission to occur, not only a high level of the expression of the *lux*CDABE genes is required to form bacterial luciferase and fatty acid reductase, but also the synthesis of the substrates of bacterial luciferase is essential to maintain the light emission over long time periods. Hence, the emission of light by luminous bacteria is highly dependent on the growth and the environment.

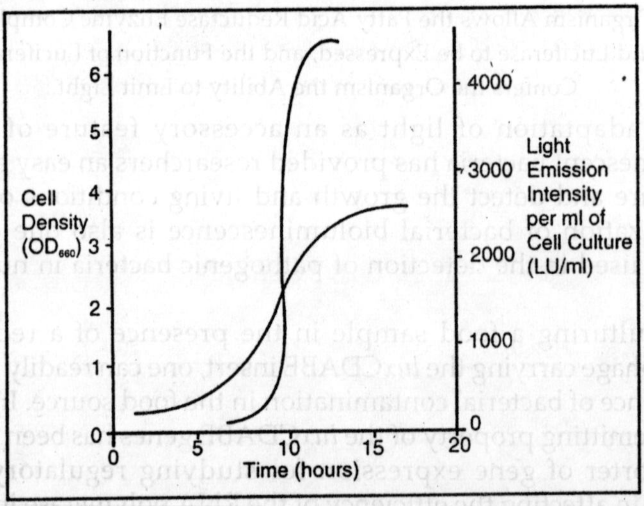

Fig. The Superimposition of the Growth Curve of Luminous Bacteria (Colored in black), and the Glowing Intensity from the Cell Culture Measured Along the Growth (Colored in Light Blue) Indicates that the Mid-log phase of the Growth of Luminous Bacteria is the "Switch Point" where "the Light is Turned on".

In the laboratory set-up luminous bacteria growing in liquid media at low cell density give off a minimal amount of light because of the dormancy of the expression of *lux*CDABE genes and the deficiency in the level of substrate for the bacterial luciferase reaction. From the mid to the end of the exponential growth period, the intensity of the light emission rises dramatically as the result of the rapid accumulation of the synthesized substrates and enzymes from the activation of the expression of the *lux*CDABE genes.

The increase in the expression of the *lux*CDABE genes occurs through a mechanism analogous to an autocatalytic chemical reaction, in which the products of the reaction also catalyze the reaction from which they were derived, leading to the generation of more products and therefore more catalysts.

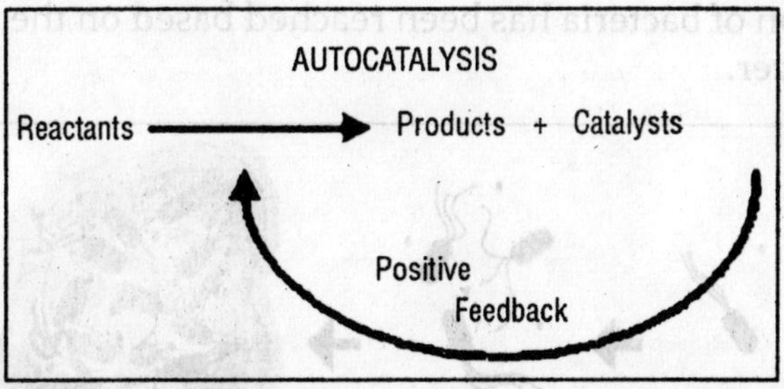

Fig. Autocatalysis is the Catalysis of the Reaction by One of the Reaction Product(s), which Exponentially and Automatically Increases the Reaction Rate as the Reaction Roceeds.

In bacterial bioluminescence the catalysts that induce the expression of the *lux*CDABE gene are regulatory proteins, and a small chemical compound, which is commonly called the autoinducer. The autoinducer is a small metabolic product, which freely diffuses across the cellular membrane, and is excreted to the extracellular environment during the early stage of cell growth.

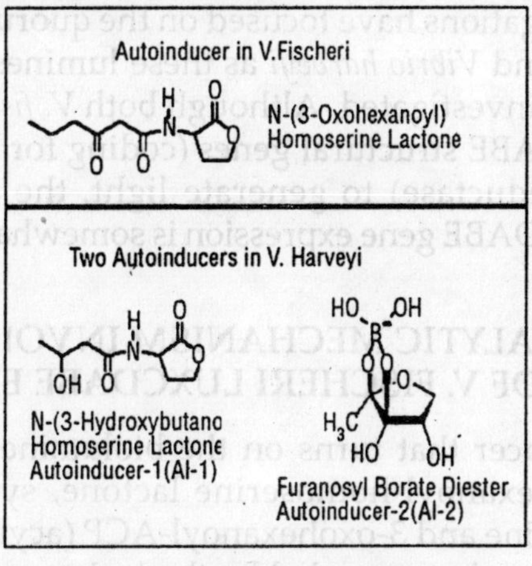

Fig. Chemical Structures of the Autoinducers of Luminous Bacteria

It is known that marine luminescent bacteria that live free in the ocean do not emit light, while bacteria living as symbionts in marine organisms and/or in a localized, confined environment emit a high level of light. Therefore, in order for the light emission to occur, luminous bacteria have to grow in a confined, nutrition-rich environment, in which the autoinducer can be accumulated.

When the concentration of the autoinducer increases to a certain level in the extracellular environment, the autoinducers will activate the luminous system of luminous bacteria. This process has been called "quorum sensing" as the bacteria literally sense when a sufficient concentration of bacteria has been reached based on the accumulation of autoinducer.

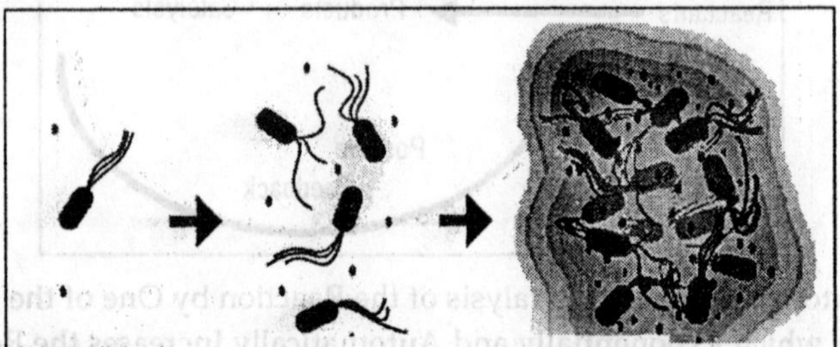

Fig. The Confined Environment, in which the Produced Autoinducers can Accumulate, is the key to the Activation of the Quorum sensing Mechanism.

Most investigations have focused on the quorum sensing systems of *Vibrio fischeri* and *Vibrio harveyi* as these luminescent bacteria have been extensively investigated. Although both *V. fischeri* and *V. harveyi* utilize the *lux*CDABE structural genes (coding for bacterial luciferase and fatty acid reductase) to generate light, the mechanism of the activation of *lux*CDABE gene expression is somewhat different between the two species.

THE AUTOCATALYTIC MECHANISM INVOLVED IN THE REGULATION OF V. FISCHERI LUXCDABE EXPRESSION

The autoinducer that turns on the bioluminescent system of *V. fischeri* is 3-oxohexanoyl-homoserine lactone, synthesized from S-adenosylmethionine and 3-oxohexanoyl-ACP (acyl carrier protein) by the autoinducer synthetase coded by the *lux*I gene.

When a critical concentration of 3-oxohexanoyl-homoserine lactone is established in the extracellular environment , autoinducer starts to accumulate inside the luminous bacteria and interacts with the regulatory protein coded by *lux*R.

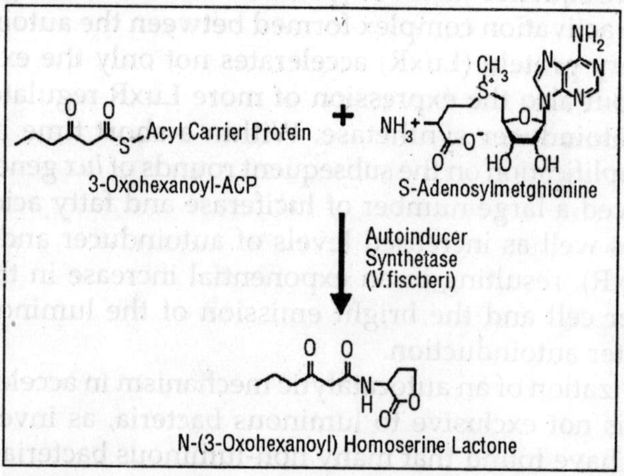

Fig. The Amide Linkage Formation from 3-Oxohexanoyl and Methionine Groups, and the Lactonization of the Methionine Side Chain are the Two Reaction Steps Involved in the Formation of the Homoserine Lactone Autoinducers.

One notable feature about the bioluminescence system in *V. fischeri* is the organization of the regulatory *lux* genes in controlling the induction of bacterial bioluminescence, particularly the bi-directional RNA polymerase promoter binding site between the oppositely oriented *lux*R operon and the contiguously linked, polycistronic operon containing the *lux*ICDABE genes.

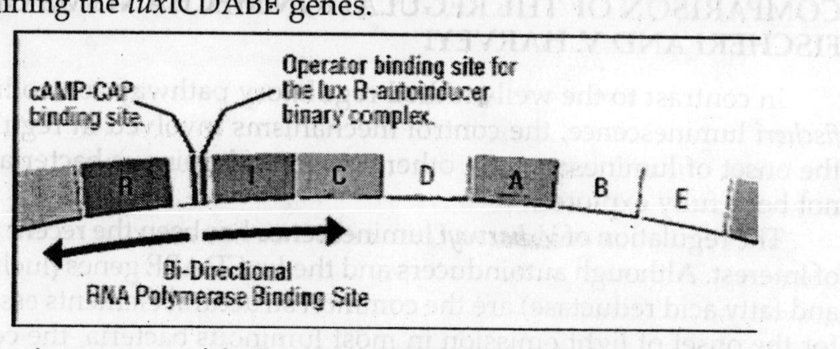

Fig. The Arrangement of the Regulatory Elements in the Upstream Region of the *lux* Structural Genes in *V. fischeri.*

Binding of the activation complex (autoinducer-LuxR) to the operator site in the regulatory region between the *lux*R and *lux*ICDABE operons along with a cAMP-CRP (cAMP-cAMP receptor protein) complex at a CRP binding site, leads to RNA polymerase being recruited to the bi-directional promoter site to transcribe the downstream *lux* genes.

As a consequence of the organization of these *lux* genes in the genome, the activation complex formed between the autoinducer and the regulatory protein (LuxR) accelerates not only the expression of *lux*CDABE but also the expression of more LuxR regulatory protein and LuxI autoinducer synthetase. Within a short time interval the feedback amplification on the subsequent rounds of *lux* gene expression have produced a large number of luciferase and fatty acid reductase molecules as well as increased levels of autoinducer and regulatory protein (LuxR), resulting in an exponential increase in the emission intensity per cell and the bright emission of the luminous bacteria observed after autoinduction.

The utilization of an autocatalytic mechanism in accelerating gene expression is not exclusive to luminous bacteria, as investigators in other fields have found that many non-luminous bacteria use similar quorum-sensing mechanisms to orchestrate the onset and enhancement of gene expression.

While the activation and the *lux*CDABE gene expression in *V. fischeri* have been well studied, the mechanism by which the luminescence appears to be repressed at very high cell densities is not well understood. This decrease in light intensity may reflect a limitation in substrate availability; alternatively, the autoinducer-LuxR activation complex may bind to a low-affinity operator-like binding site causing repression.

COMPARISON OF THE REGULATION MECHANISMS OF V. FISCHERI AND V. HARVEYI

In contrast to the well-studied regulatory pathway controlling *V. fischeri* luminescence, the control mechanisms involved in regulating the onset of luminescence in other species of luminous bacteria have not been fully explored.

The regulation of *V. harveyi* luminescence has been the recent target of interest. Although autoinducers and the *lux*CDABE genes (luciferase and fatty acid reductase) are the common structural elements essential for the onset of light emission in most luminous bacteria, the control mechanism by which the level of the expression of *lux*CDABE is regulated in *V. harveyi* is extremely complex compared to that of *V. fischeri*. In contrast to the *lux* I/R pair of regulatory genes in directing *lux*CDABE expression in *V. fischeri*, a multitude of at least eight regulatory *lux* genes are involved in signal transduction in controlling the onset of *V. harveyi* luminescence. However, the signal transduction in *V. harveyi* from the autoinducer quorum sensor in the extracellular environment to the operon in the cell is functionally homologous to the *lux*I/R of *V. fischeri* in luminescence activation.

The purpose of the partitioning of the integrated function over various regulatory components may be due to the coupling of the luminescence regulation to one or more metabolic pathways and fine-tuning the level of luminescence emission in *V. harveyi* in response to nutritional signals.

It therefore is necessary to elucidate the function of each regulatory element involved in the signal relay system to determine the potential interactions of these regulatory components with other cellular metabolic pathways and their contributions to the modulation of the level of luminescence emission. Consequently, the partitioned functions of the regulatory components in *V. harveyi* in relation to the integrated *lux* I/R in *V. fischeri* will be briefly described here.

In contrast to the closely linked and contiguous *lux*R and *lux*ICDABE operons in the *V. fischeri* genome, all the regulatory *lux* genes involved in controlling the *lux*CDABE expression are scattered over the *V. harveyi* genome. In addition, there are two signaling transduction pathways, responding to two different autoinducers in *V. harveyi* luminescence regulation.

The autoinducer, N-(3-hydroxybutanoyl)homoserine lactone, is produced by the autoinducer synthetase encoded by *lux*M and the other autoinducer, a furanosyl borate diester, is synthesized by the *lux*S autoinducer synthetase.

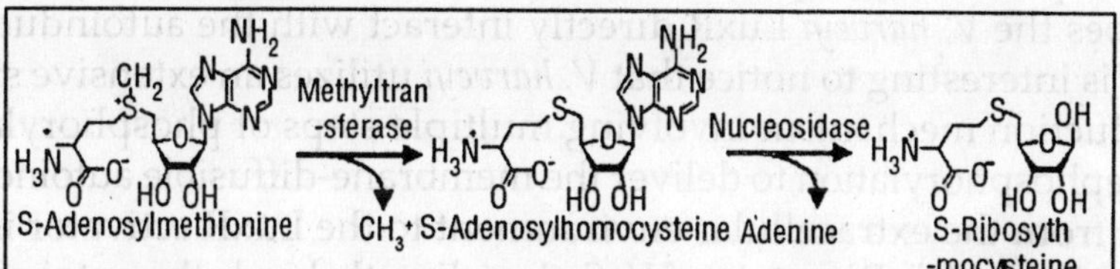

Fig. The Reaction Mechanisms Involved in the Biosynthesis of Furanosyl Borate Diester, the Autoinducer-2 in *V. harveyi*.

Although the *lux*I synthetase of *V. fischeri* synthesizes a homoserine lactone derivative that is similar to that produced by the LuxM synthetase of *V. harveyi*, neither LuxM nor LuxS share sequence homology to autoinducer synthetase (LuxI) of *V. fischeri*.

In contrast to the formation of the transcriptional activator complex (LuxI autoinducer and LuxR regulatory protein), which directly acts on the *lux*CDABE operon to activate transcription in *V. fischeri*, the autoinducers excreted into the extracellular environment by *V. harveyi* bind to corresponding autoinducer receptors (LuxM autoinducer binds the extracellular domain of the LuxN transmembrane protein; LuxS autoinducer binds the extracellular LuxP subunit of the LuxP/Q transmembrane protein complex) on membrane proteins, which have autoinducer receptor domains in the extracellular environment and kinase/phosphatase domains in the intracellular environment.

The binding of autoinducers to the corresponding receptor domains then causes a conformational change in the intracellular domains, activating the intracellular phosphatase activity of LuxN and LuxP/Q transmembrane signal mediators. From this point on, the signal converges, and both the phosphatase domains of LuxN and LuxP/Q dephosphorylate the LuxU regulatory protein, which then dephosphorylates LuxO, removing the repression from the *lux*CDABE genes. LuxO is a sigma-54 regulator and is believed to stimulate synthesis of a repressor of the *lux* operon and/or *lux*R. Once the LuxO regulatory protein is dephosphorylated, the repression on *lux*CDABE expression is relieved, leaving LuxR as the activator of the *lux*CDABE operon. However, it is important to emphasize that *V. harveyi* LuxR does not posses any sequence or structural homology to *V. fischeri* LuxR nor does the *V. harveyi* LuxR directly interact with the autoinducer.

It is interesting to notice that *V. harveyi* utilizes an extensive signal transduction mechanism involving multiple steps of phosphorylation and dephosphorylation to deliver the membrane-diffusible autoinducer signal from the extracellular environment to the LuxR activator in the cell, while the LuxR protein of *V. fischeri* directly binds the autoinducer to activate *lux* CDABE expression.

As both the phosphorylation and dephosphorylation are integral activities of various signal transductions pathways involved in controlling cellular growth, the transfer of the quorum-sensing signal by a phospho-transduction system to the *lux* operon provides an excellent opportunity to integrate metabolic energy with the quorum sensing regulation in luminous bacteria. The phenomena of bioluminescence in nature have always been a focus of human attention. Prior to the identification of luminous organisms, the presence of even a dim luminous glow in the dark frightened countless numbers of individuals who perceived the glimmer as ghosts or supernatural spirits.

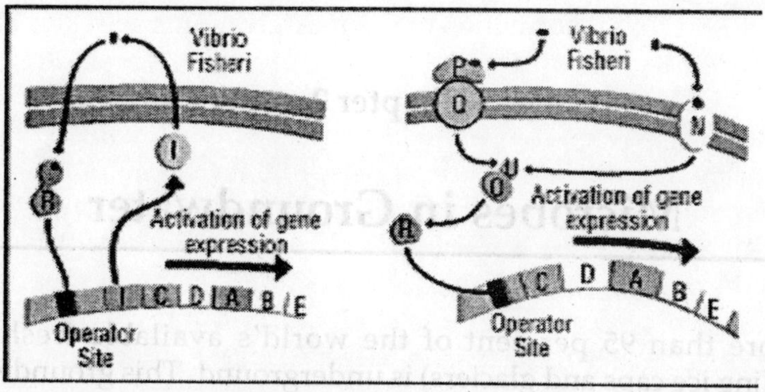

Fig. The Comparison of the Different Mechanisms Involved in the Regulation of the Expression of the *lux* Structural Genes in *V. fischeri* and *V. harveyi*.

However, the curiosity of scientists to solve the mysteries of nature led to the identification of many luminous organisms responsible for the light emission in different environmental settings. The isolation of luminous bacteria from marine and terrestrial environments and the characterization of their properties have resulted in the identification of a number of luminous bacterial species classified into several genera.

All luminescent bacteria utilize $FMNH_2$, O_2 and long fatty aldehyde as substrates for the bioluminescence reaction catalyzed by luciferase (LuxAB) with the fatty acid reductase complex (LuxCDE) synthesizing the long chain aldehyde substrate of tetradecanal. A NAD(P)H-dependent FMN reductase found in most if not all bacteria generates the $FMNH_2$ substrate. Consequently, only the five genes in the *lux* operon, *lux*CDABE, are needed to produce light emission even in bacteria that normally do not emit light and thus provide the opportunity to utilize the bacterial *lux* system as a light emitting sensor in many bacteria. The regulation of the induction of expression of the *lux*CDABE genes in luminescent bacteria at high cell density has led to the development of molecular prototypes for quorum sensing, an important new regulatory mechanism signaling bacterial crowding and now found controlling key secondary metabolic pathways in many nonluminescent bacteria. Advances in understanding the molecular mechanisms, structures and regulation of the *lux* enzymes and genes of luminescent bacteria has answered many of the questions but why bacteria utilize the energy of the cell to emit light in this exergonic process and how the luminescent system interfaces at the molecular level with the metabolic and energy supplying pathways in the bacteria still remains a mystery.

Chapter 2

Microbes in Groundwater

More than 95 per cent of the world's available fresh water (excluding ice caps and glaciers) is underground. This groundwater is valuable as a source of drinking water for most communities in the world, especially small ones.

In the United States, about 15 million private wells serve fewer than 25 people each. In addition, about 92 per cent of the 168,000 water systems serving 25 people or more get most or all of their drinking water from groundwater. The common microbes in groundwater (bacteria, protozoa, and viruses), primarily from a public health point of view.

COMMONLY OCCURRING MICROBES

Groundwater near the land surface usually teems with microbial life. Bacteria, which are microscopic single-celled organisms that lack a true nucleus and normally have a cell wall, are far more numerous than any other organism in the soil and groundwater. Near the surface where plant roots are abundant, there may be 100 million to 1 billion bacteria per gram of dry soil.

These values decrease dramatically with soil depth below the root zone, with densities depending on the amount of nutrients (food) and water available. Although concentrations of microbes below the root zone are lower than in the root zone itself, as many as 10 to 100 million bacteria per gram of aquifer material may be present. Bacteria have been found in core samples from a depth of 2.8 kilometers (1.7 miles) below the Earth's surface, and at a depth of 3.2 kilometers in South African gold mines.

Protozoa, single-celled organisms that have a nucleus but lack a cell wall, also are common in groundwater. Protozoa typically are much larger than bacteria, and many types in groundwater feed on bacteria. Molds and other fungi also are common in groundwater near the land surface, where plenty of oxygen is present.

These microbes are larger than bacteria and, unlike the protozoa, have a cell wall and often grow in long filaments. Most fungi feed on dead or decaying material. Algae may even be present despite the absence of sunlight. Similar to the bacteria, the numbers of protozoa, fungi, and algae decrease with depth.

A large variety of viruses probably are also common in groundwater. Viruses are much smaller than other microbes and cannot be seen with ordinary microscopes. They do not have a nucleus or cell wall, and can multiply only within the cells of larger organisms. The natural viruses in groundwater are able to reproduce only by infecting and usually killing the bacteria and other larger microbes present.

Microbial Habitats

Microbes below Earth's surface must compete vigorously with each other for limited food and space. Bacteria that are not adapted to this habitat cannot survive long. Normally, nutrients are scarce in this environment, and thus microbes in groundwater grow slowly.

Because the microbes in groundwater and other environmental habitats are well adapted to life in these habitats, few of them have the ability or the need to infect and cause illness in humans and other animals. However, a few environmental bacteria and protozoa may cause disease in humans by chance under certain circumstances, especially in individuals with a weakened immune system or major breaks in the skin that allow microbes to enter.

Many naturally-occurring bacteria below the surface may serve a beneficial role by destroying human-made toxic chemicals that migrate from the surface to groundwater. Some toxic chemicals, such as gasoline, are easy for subsurface bacteria to break down, but others, such as dry-cleaning fluid and some pesticides, are destroyed only slowly, if at all.

Disease-Causing Pathways

The groundwater in a drinking-water well may contain a wide variety of microbes without presenting a public health risk. However, groundwater in some areas becomes contaminated by the fecal material of humans and other animals. This is a cause for concern because fecal material may contain pathogenic (disease-causing) microbes that can infect the intestinal tract of humans. Fecal pathogens may be bacterial, viral, or protozoan. Water containing fecal material may seep into the groundwater from the land surface or from underground sources of contamination. Major surface sources include:

- Wastewater and biosolids from sewage treatment facilities that have been applied to land as a soil conditioner.

- Seepage from shallow artificial ponds (lagoons) used for processing sewage.
- Seepage from contaminated lakes and other surface-water bodies.
- Urban runoff.
- Feces from cattle and other livestock operations.
- Improperly constructed sanitary landfills where trash and garbage are disposed.

Fecal contamination also can reach the groundwater from underground sources, such as improperly functioning septic tank systems, underground reservoirs for liquid household sewage (cesspools), or leaking underground sewer lines. About 25 million septic tanks exist in the United States.

PATHOGEN MOVEMENT AND PERSISTENCE

Most pathogens from fecal matter remain either near the surface, or near the point of origin in the case of an underground source. However, where conditions are favorable, some pathogens travel along with the water flow though pores (tiny openings) in the surrounding soil and rock, and may enter groundwater.

Whether a pathogen reaches the groundwater, especially in the area of a well intake, depends on how strongly it is retained by the soil and how long it survives. The extent to which a pathogen is retained by the soil depends on:

- Soil characteristics, such as the number and size of pores and their interconnections, and how strongly the pathogen adsorbs (binds) to the soil particles;
- Environmental factors, especially the amount of rainfall; and
- The characteristics of an organism, including its size and its tendency to attach to other organisms and organic debris.

Rapid pathogen transport may occur in aquifers that have large pores (such as in a gravel-dominated aquifer), as well as in fractured rock, cave systems, and sinkholes.

Its presence in groundwater indicates that surface water is gaining access to the aquifer, either naturally or induced through pumping. *Cryptosporidium* may cause serious intestinal illness.

Pathogen survival depends on a number of factors, including soil moisture content (longer survival when moist), soil temperature (longer survival at low temperatures), soil acidity (shorter survival in acid soils), amount of nutrients (longer survival of bacterial pathogen if high), and the activity of microbes that normally live in the soil.

Normal soil bacteria are fully adapted to their environment and should easily outcompete bacterial pathogens for available nutrients. Some will also release one or more chemicals (toxins) that may kill the pathogens.

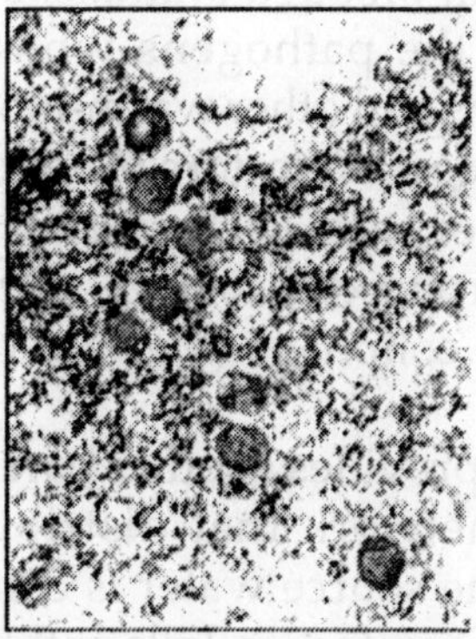

Fig. *Cryptosporidium* is a Protozoan that Occurs Naturally in Surface Waters Such as Streams and Lakes

In most cases, only a small portion of the original pathogen density will reach a well intake unless a rapid pathway exists (such as fractured rock). Despite this, most reported waterborne disease outbreaks reported in the United States each year are associated with groundwater. Between 1971 and 2000, almost 60 per cent of the approximately 700 reported outbreaks, with more than 90,000 cases of illness, were associated with water systems using groundwater.

PROTECTING PUBLIC HEALTH

How can the public be protected from fecally contaminated groundwaters? Drinking-water supplies using groundwater can be protected from fecal contamination of the groundwater in several ways. Most importantly, suppliers can ensure that their well meets the state well-construction code.

They can identify any sources of fecal contamination near the well intake and take proper corrective measures. Suppliers can also have an expert conduct a periodic on-site inspection, including an assessment of how vulnerable the source water is to nearby sources of fecal contamination.

Another means by which a water supply can be protected is to periodically test its source-water quality. To test for all possible pathogens, however, is not practical; the wide variety of potential waterborne pathogens and the limitations of analytical methods would make testing for all such pathogens extremely difficult, time-consuming, and expensive. Instead, inexpensive and easily measured indicators of fecal contamination are used rather than testing for specific pathogens.

These include microbes that are extremely common in the gut of humans and other warm-blooded animals, but not elsewhere. If such indicators are absent, the pathogens whose normal habitat is the intestinal tract of humans and other warm-blooded animals should also be absent. Among the fecal indicators used for environmental waters, including groundwater, are *Escherichia coli* (a bacterium), fecal streptococci (a small group of bacteria), enterococci (very similar to fecal streptococci), and coliphage (a group of viruses that infect and kill *E. coli*).

Water disinfection provides another barrier to pathogen entry and survival in the pipe network that distributes water (distribution system), especially if the source water is contaminated or vulnerable to fecal contamination. Disinfection kills pathogens and other microbes. Commonly used disinfectants for groundwaters include chlorine and ultraviolet light.

The U.S. Environmental Protection Agency develops regulations that apply to every water supplier that serves at least 25 people for at least 60 days during a year (defined as a public water system). One regulation that protects groundwater systems from pathogens is the Total Coliform Rule (TCR). The TCR requires every public water system to monitor its tap water

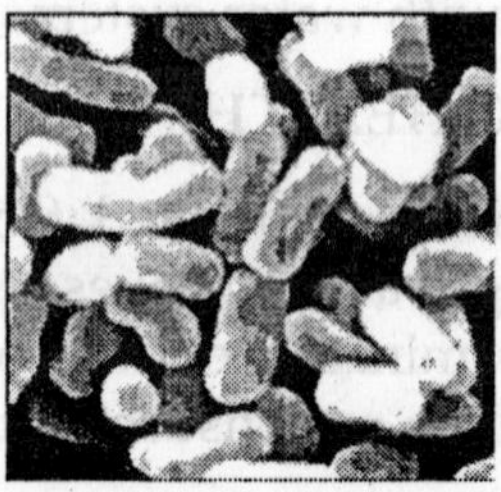

Escherichia coli is a bacteria found in the intestinal tract of warm-blooded animals. Its presence in groundwater indicates that a pathway exists between the land surface and the aquifer, and may indicate that the water is not safe to drink.

For a group of closely related bacteria, called total coliforms, at a frequency that depends on the number of people a system serves. While total coliforms are not a fecal indicator, they are used to detect problems in water treatment and in the distribution system. If any sample is total coliform positive, the system must then test for a fecal indicator (normally *E. coli*).

Another regulation, the Surface Water Treatment Rule (SWTR), in addition to surface waters, also covers groundwater systems that are directly under the influence of surface waters. Among other requirements, the SWTR requires all such systems to disinfect. An important element in both the TCR and the SWTR, as amended, is a requirement for systems to have a periodic on-site inspection (sanitary survey) by a state official or a state-approved agent.

As of 2002, the Environmental Protection Agency was developing another regulation, the Groundwater Rule, which will require groundwater systems to determine whether the source water is fecally contaminated or is vulnerable to such contamination. If so, the system will be required to take corrective measures.

DRINKING WATER TREATMENT

Public water systems vary in the treatment of water that is delivered to the consumer, depending on whether the system uses a groundwater source or a surface-water source. Some public water systems, primarily those that utilize groundwater wells, deliver untreated water directly to the customer's tap. Groundwater is usually less susceptible to contamination than surface water, and therefore requires little or no treatment.

Many large public water systems, however, draw their water from surface sources such as lakes, rivers, and streams, which are vulnerable to many types of contamination. Hence, a multiple-barrier approach to water treatment is most effective in producing drinking water that is free of health risks, meets regulatory requirements, and is palatable (acceptable in taste) to the consumer.

Groundwater

Although groundwater generally is less likely to be contaminated than surface water, it nonetheless may require contaminant removal. Possible groundwater contaminants include naturally occurring inorganic chemicals such as arsenic, fluoride, and nitrate, as well as human-made chemicals (from solvents, fuels and pesticides, for example) that have found their way into the aquifer.

Other contaminants that present aesthetic concerns are calcium and magnesium (which produce hardness); iron and manganese (which cause staining of laundry and plumbing fixtures); and hydrogen sulfide gas (which produces the unforgettable "rotten-egg" smell). The word "contaminant" includes any substance in water, including those that are not harmful, even though popular usage often refers only to harmful pollutants.

Hard water is usually treated with the water "softening" method, which utilizes a resin-coated media that exchanges its ion (usually sodium) for the calcium and magnesium ions that are responsible for hardness. The watersoftening media must be regenerated periodically with a brine (very salty) solution.

Even if the water is not hard, inorganic chemicals such as iron and manganese can cause problems that require treatment. Iron and manganese can be oxidized, or brought out of solution, by chemicals such as chlorine.

Settling tanks at a water treatment plant allow coagulated particles to settle to the bottom of the tank. The settled material is periodically drained and collected for eventual disposal in a landfill or potassium permanganate. The oxidized metals can then be filtered out as the water passes through a sand media. A zeolite sand, commonly known as manganese greensand, is usually used for this filtration.

Some groundwaters have taste and odor problems that require treatment in order for the water to be palatable to the user. Chlorine and ozone are effective in reducing hydrogen sulfide odors, and simple aeration can rid the water of some volatile organic compounds, such as solvents or petroleum byproducts.

SURFACE WATER

Many large cities are served by surface-water sources; most of these surface waters are exposed to a variety of contaminants. Common contaminant sources are untreated sewage and runoff from fertilized fields, parking lots, or unprotected watersheds. The primary contaminants of concern are microbes, including *Giardia* and *Cryptosporidium.* Effective surface water treatment therefore relies on the multiple-barrier approach, involving a sequence of processes from among the following options.

Pre-Treatment[sh]

Simple screens over intake pipes in lakes or rivers can prevent large debris, such as leaves, sticks, or small fish from entering the treatment plant. Waters that exhibit seasonal changes in turbidity or algae growth may require an oxidant such as chlorine, potassium permanganate, or ozone for pre-treatment.

Chemical Feed and Rapid Mix.

Coagulants such as alum, ferric chloride, or synthetic polymers are added to the raw water and mixed rapidly. Chemicals to adjust the pH may also be added at this point. The effectiveness of further treatment depends partly on raw water chemical characteristics such as pH, temperature, and alkalinity.

Coagulation, Flocculation, and Sedimentation

The coagulant chemicals, by the process of coagulation, neutralize the electrical charges of the sediment particles, allowing the binding together of small particles into larger particles that can then be settled or filtered. This process is known as flocculation.

Small filters made of pleated paper, carbon, or synthetic resins often are used by consumers. Pleated paper filters can remove sand, dirt, and some iron. Carbon filters are effective against chlorine and organic chemicals such as herbicides and pesticides, and can also remove objectionable tastes and odors.

Some manufacturers of resin-type filters claim a high removal of lead or other inorganic chemicals.

The sedimentation process then allows the larger "floc" particles to settle in a basin. The cleaner water at the top of the sedimentation basin flows onto the filter.

Filtration

The filter media is usually a mixture of carefully graded and specified gravels and sands, specifically designed for particle retention. These removed particles include turbidity and biological contaminants such as *Giardia* and *Cryptosporidium*. Inorganic chemicals such as arsenic and mercury are also removed by proper coagulation and filtration.

Disinfection

This process is designed to kill disease-causing microorganisms such as bacteria, viruses, and protozoa. Chlorine is most commonly used for this purpose, although ozone and ultraviolet light are increasingly being used in public water systems. The single-celled protozoa (*Giardia* and *Cryptosporidium*), which can cause severe intestinal illness, are more resistant to traditional disinfectants than bacteria or viruses, and the disinfection process must therefore be closely monitored.

OTHER TREATMENT TECHNOLOGIES

A number of other treatment technologies are employed both by the public and private sector, depending on the needs and desires of the water users. Granular-activated carbon is used for the elimination of objectionable tastes and odors, as well as for removal of organic chemicals such as pesticides and herbicides.

Sodium carbonate, commonly known as soda ash, can be used to raise the pH of the water to make it less corrosive to pipes and plumbing fixtures. Some public water systems add small amounts of fluoride in order to aid in the prevention of dental cavities in children.

In addition, many small water treatment units, designed to fit under the kitchen sink or in any other convenient location, are used by individual homeowners regardless of the water source. Cartridge filters, membrane filters, and reverse osmosis units are all able to filter out most contaminants that cause health risks and aesthetic concerns. Water softeners are often necessary to treat hard water. Small ozone or ultraviolet-light disinfection units are becoming popular with homeowners who depend on private water supplies of questionable quality. Many different types of water treatment units are designed for the individual homeowner. When considering the purchase of such a unit, the homeowner should have the water tested to determine exactly what type of contaminants are in the water, and what, if anything, needs to be removed.

No one filter works for all contaminants. A filter that removes iron and manganese, for example, will probably not remove herbicides or pesticides. It is also important to remember that all treatment units require regular maintenance to keep them functioning properly and to prevent the buildup and possible breakthrough of the unwanted contaminants.

Ecology, Fresh-Water

Streams, lakes, and wetlands differ profoundly from one another in the conditions they provide as habitats for biological communities. Fundamental characteristics of standing water (a lentic system) or flowing water (a lotic system), the dynamics of its interaction with adjacent land and vegetation, and seasonal fluctuations in water conditions determine characteristic biological assemblages.

STREAMS

The most apparent feature of streams is flowing water. The geomorphology and topography through which streams flow, their channel steepness, the variability in stream-bottom substrates, and the availability of woody debris are among the natural controls that influence flow patterns. The diversity in physical structure and flow among riffles, pools, and glides is reflected in equally diverse aquatic communities.

A flowing-water habitat is known as a lotic ecosystem. A lentic system, such as a lake, is predominantly nonflowing.

In flat valley floodplains, streams naturally meander, or flow in a winding path. In these systems, flooding is an important natural process: nutrient-rich sediments are deposited onto the floodplain, and young fish are protected in quieter areas away from high flow.

Terrestrial (land-based) invertebrates become important sources of food during these times. Under the stream and beneath floodplain soils, water flows in the subsurface in the hyporheic zone. Waters are cooled, nutrients are exchanged, and invertebrates are specially adapted to live in this subterranean habitat.

A river basin consists of its entire upstream network: its watershed is the landscape, including the stream network and the land it traverses. Riparian (streamside) zones make up the transition between stream and terrestrial environment, encompassing the area where vegetation and streams interact. Riparian plants provide shade, cool the stream, retain soils, and filter nutrients. Fallen leaves decompose and provide nutrients for stream microorganisms and invertebrates.

Changing habitats along the longitudinal stream gradient, from its headwaters to downstream reaches, can be thought of as a stream continuum. Energy possessed by the stream and aquatic communities tend to follow this continuum in predictable ways. Leafy inputs are most important in headwaters, whereas algae are more important as a food base in wider, more sunlit reaches. Nutrients released from streambanks and within the stream are taken up by algae in a downstream spiral of repeated uptake and release.

Macroinvertebrates in the stream continuum are classified into functional feeding groups, depending on how they obtain food. In headwaters, shredding invertebrates, dependent on decomposing leaves, are common.

As streams open up, scraping macroinvertebrates that scoop algae off rocks are more abundant. Collector invertebrates either gather or filter out small edible particles everywhere, but in wide, deep rivers they dominate with plankton. As adults many stream invertebrates move into the terrestrial environment.

In stream food webs, primary production (the foundation of food chains) comes from riparian inputs and algae, and most invertebrates are herbivores or detritivores.

Top predators are primarily fish but also include birds, amphibians, and humans. Invertebrate predators are relatively few. Fish move easily between habitats, and many feed on both stream and riparian invertebrates. A few birds spear large invertebrates that avoid fish predation. Other birds and bats capture adult aquatic insects that emerge from the stream.

LAKES

Lakes are characterized by basin shape, volume, and depth. Shallow edges where rooted vegetation persists are nursery grounds for fish and diverse invertebrates.

The open-water (pelagic) zone is classified into layers, depending on the degree of mixing that occurs. Water changes density with temperature, causing lakes to become layered, or stratified, into temperature zones.

Consequently, significant energy is needed to mix a lake's thermal layers. Deep lakes are stratified into a warmer, upper layer that mixes readily (epilimnion); a middle layer of quickly changing temperatures (metalimnion); and a deep, cool bottom layer that mixes infrequently during the year (hypolimnion). Wind easily mixes shallow lakes, so these layers either do not persist or do not develop.

Phytoplankton, predominantly algae, form the base of a lake's food chain. These primary producers fall into five major categories. These major categories also are represented in streams.

A community of microscopic zooplankton lives in a lake's upper layer, feeding on single-celled or small colonial algae. In clear, relatively unproductive lakes, zooplankton consume much of the algae. Very productive lakes are much less clear, with abundant algae in blooms of filamentous forms.

COMMON PHYTOPLANKTON	
Group	Characteristics
Diatoms (*Bacillariophyta*)	Primary pigments are chlorophyll and carotenes so they appear yellow-brown in color, have cell walls made from silica, often have early season dominance, cell walls may persist for thousands of years in sediments where they are useful for determining historic water characteristics.
Green Algae (*Chlorophyta*)	Primary pigment is chlorophyll so they appear bright green, population
growth often follows spring diatoms, are important food for zooplankton, occasionally cause nuisance blooms especially when nitrogen concentrations in the lake are high.	
Golden Algae (*Chrysophyta*)	Primary pigments are chlorophyll, carotene and xanthophylls, appear
yellow-brown in color, optimize their growth during the relatively cool waters of late winter/early spring, taste and odor problems are often associated with golden algae, may supplement photosynthesis by capturing food particles in the water.	
Blue-Green Algae (*Cyanobacteria*)	Primary pigments are chlorophyll, carotene, phycobilins, and xanthophylls, appear blue-green, no chloroplasts-color is distributed evenly throughout cell, are similar biologically to bacteria, often form filamentous or spherical colonies, can form nuisance blooms, often dominate summer plankton of productive lakes, most can regulate their buoyancy, many are nitrogen-fixers, some produce potent toxins.

Dinoflagellates *(Dinophyta)*	Appear dark green to brown to almost black in color, are free-swimming, have two flagella, have forward-spiraling swimming motion, blooms are associated with organic pollution, may cause taste and odor problems in water, marine dinoflagellates are responsible for "red tides."

Water chemistry plays an important role in lake dynamics, as nutrients influence algal productivity and higher trophic levels. In deep lakes, the bottom layer has little oxygen when it is not mixing, and few organisms survive there. Similarly, very salty lakes contain only a few highly specialized zooplankton.

When fish occur in lakes, they are the dominant predators. Fish are visual predators and select prey on the basis of size. They consume the largest zooplankton, and remaining zooplankton will be small. A trophic cascade occurs as the top predators (fish) depress the next trophic level (zooplankton), releasing high production in a third level (algae). In fishless lakes, invertebrate predators dominate, and large herbivorous zooplankton escape predation.

WETLANDS

Wetlands generally are more shallow than lakes and have standing or flowing water above, near, or below the surface. They vary in size, from puddles to thousands of hectares (a hectare is 10,000 square meters). Because water is not necessarily present year-round, characteristic waterlogged soils or plants help identify their location and extent. Wetlands often are intermediate habitats, representing a transition between aquatic and terrestrial habitats. Their transitional position makes them key exporters of organic materials and sinks for nutrients. Persistent, shallow standing water forms a permanent wetland. When soils are saturated for only part of the year, temporary or ephemeral wetlands are created.

Wetland communities reflect the duration and depth of standing water. Plants vary specifically in requirements for annual wetness. Similarly, invertebrates differ in intervals required to mature, eventually migrating or becoming dormant in the dry season. Some amphibians use only a seasonal aquatic habitat before migrating; others require continuous wetlands.

Ephemeral plant and invertebrate resources will attract migratory waterfowl, but resident birds and fish require year-round water. Thus, a variety of wetlands support diverse plant and animal communities.

Human Influences

Habitat loss due to human activities has severely affected fresh-water ecosystems. Because of filling and draining activities more than half (53%) of wetlands in the continental United States were lost between 1780 and 1980. Dams, channelization, wood removal, and flow alteration have reduced stream habitat significantly.

Profound changes from accidental and intentional introduction of nonnative plant and animal species have altered aquatic communities and destroyed habitats. Acid rain and other consequences of air and water pollution have severely damaged lakes, streams, and wetlands.

Cumulative effects of these activities often result in declining native flora and fauna. Although scientists and decisionmakers recognize many alterations caused by humans, they have just begun to develop scientific and political understandings for fresh-water restoration.

ALGAL BLOOMS IN FRESH WATER

Aquatic ecologists are concerned with blooms (very high cell densities) of algae in reservoirs, lakes, and streams because their occurrence can have ecological, aesthetic, and human health impacts.

In waterbodies used for water supply, algal blooms can cause physical problems (e.g., clogging screens) or can cause taste and odor problems in waters used for drinking. Blooms involving toxin-producing species can pose serious threats to animals and humans.

Algae in Aquatic Ecosystems

The term "algae" is generally used to refer to a wide variety of different and dissimilar photosynthetic organisms, generally microscopic. Depending on the species, algae can inhabit fresh or salt water.

In modern taxonomic systems, algae are usually assigned to one of six divisions. The misnamed blue-green algae are often grouped with algae because of the chloroplasts contained within the cells. However, these organisms are actually photosynthetic bacteria assigned to the group cyanobacteria.

Fresh-water algae, also called phytoplankton, very in shape and colour, and are found in a large range of habitats, such as ponds, lakes, reservoirs, and streams. They are a natural and essential part of the ecosystem. In these habitats, the phytoplankton are the base of the aquatic food chain. Small fresh-water crustaceans and other small animals consume the phytoplankton and in turn are consumed by larger animals.

Bloom Occurrences and Impact

Under certain conditions, several species of true algae as well as the cyanobacteria are capable of causing various nuisance effects in fresh water, such as excessive accumulations of foams, scums, and discoloration of the water. When the numbers of algae in a lake or a river increase explosively, an algal "bloom" is the result. Lakes, ponds, and slow-moving rivers are most susceptible to blooms.

Algal blooms are natural occurrences, and may occur with regularity (e.g., every summer), depending on weather and water conditions. The likelihood of a bloom depends on local conditions and characteristics of the particular body of water. Blooms generally occur where there are high levels of nutrients present, together with the occurrence of warm, sunny, calm conditions.

However, human activity often can trigger or accelerate algal blooms. Natural sources of nutrients such as phosphorus or nitrogen compounds can be supplemented by a variety of human activities. For example, in rural areas, agricultural runoff from fields can wash fertilizers into the water. In urban areas, nutrient sources can include treated wastewaters from septic systems and sewage treatment plants, and urban stormwater runoff that carries nonpoint-source pollutants such as lawn fertilizers.

An algal bloom contributes to the natural "aging" process of a lake, and in some lakes can provide important benefits by boosting primary productivity. But in other cases, recurrent or severe blooms can cause dissolved oxygen depletion as the large numbers of dead algae decay.

In highly eutrophic (enriched) lakes, algal blooms may lead to anoxia and fish kills during the summer. In terms of human values, the odors and unattractive appearance of algal blooms can detract from the recreational value of reservoirs, lakes, and streams. Repeated blooms may cause property values of lakeside or riverside tracts to decline.

Toxic Blooms

Some algae produce toxic chemicals that pose a threat to fish, other aquatic organisms, wild and domestic animals, and humans. The toxins are released into the water when the algae die and decay.

The most common and visible nuisance algae in fresh water, and the species that are often toxic, are the cyanobacteria. A cyanobacterial bloom will form on the surface and can accumulate downwind, forming a thick scum that sometimes resembles paint floating on the water. Because these mats are blown close to shore, humans and wild and domestic animals can come into contact with the unsightly material.

Blooms of toxic species of algae and cyanobacteria can flood the water environment with the biotoxin they produce. When toxic, blooms can cause human illnesses such as gastroenteritis (if the toxin is ingested) and lung irritations (if the toxin becomes aerosolized and hence airborne). Other cyanobacterial toxins are less drastic, and cause skin irritation to people who swim through an algal bloom. Toxicity can sometimes cause severe illness and death to animals that consume the biotoxin-containing water.

Cyanobacterial toxins are known to affect bean photosynthesis when they are present in irrigation water. The toxins also can modify zooplankton communities, reduce growth of trout, and interfere with development of fish and amphibians. In some cases, toxins can be bioconcentrated by fresh-water clams.

Some algal blooms in fresh water may only be a nuisance, but others can deplete dissolved oxygen in the water or generate biotoxins that are harmful to birds, fish, and other animals. This Canada goose swims among a floating layer of heavy, but probably harmless, algal growth.

Microcystins comprise the most common group of about fifty cyanobacterial toxins. Among these toxins are ones that, if ingested in sufficient quantity, can harm the liver (hepatotoxins) or nervous system (neurotoxins). Microcystins can persist in water because they are stable in both hot and cold water.

Even boiling the water, which makes the water safe from harmful bacteria, will not destroy microcystins. As a result of this threat, the Canadian government implemented a recommended water-quality guideline of 1.5 ìg per liter of microsystin-LR (the most common hepatotoxin), and other countries will likely follow suit.

In Canada as well as the United States, there are few reports of injury and no reports of human deaths resulting from microcystins in drinking water, in large part because surface-water sources of drinking water (e.g., reservoirs, lakes, and rivers) must undergo filtration and chlorination at water utilities prior to being distributed to customers. (Cyanobacterial toxins can be removed from water only by activated charcoal filters and chlorination.)

Control Considerations

Repeated episodes of algal blooms can be an indication that a river or lake is being contaminated, or that other aspects of a lake's ecology are out of balance. While cyanobacterial blooms receive the most public and scientific attention, the excessive growth of other algae and other aquatic plants also can cause significant degradation of a lake or pond, particularly in waters receiving sewage or agricultural runoff.

Aquatic biologists and other water-quality specialists often are called to identify the causes and recommend management steps to reduce or control the problem.

However, prevention of a problem is always better than trying to fix the problem after it happens. Controlling agricultural, urban, and stormwater runoff; properly maintaining septic systems; and properly managing residential applications of fertilizers are probably the most effective measures that can be taken to help prevent human-induced fresh-water algal blooms.

BIODIVERSITY

Biodiversity describes the variety of biological organisms in a given habitat, area, or ecosystem. It includes several components involving variation in species, ecosystems, and genetics. Species diversity is perhaps what most people think of when they think of biodiversity. Species biodiversity describes the number of different species of plants, animals, fungi, protists, or bacteria within an area of interest. Although there are approximately 1.5 million named species globally, the large majority of species have yet to be studied or named, and there are likely to be as many as 10 million species on Earth.

Ecosystem diversity describes the variety of habitat types found within a given area. For example, there are numerous types of terrestrial (land-based) ecosystems in the United States, including forests, grasslands, wetlands, and deserts. There also are many types of aquatic (water-based) ecosystems, including lakes, rivers, estuaries, coral reefs, and intertidal habitats.

Genetic diversity describes genetic variation within species and represents a third component of biodiversity. Genetic variation allows species to adapt to changes in their environment. Species that are reduced to very small populations lose much of the gene pool and hence can lose much of this adaptability.

Aquatic Biodiversity

Aquatic biodiversity describes the diversity of species and ecosystems found in and around aquatic habitats such as rivers, lakes, and oceans. As with terrestrial ecosystems, aquatic biodiversity varies from region to region. Aquatic biodiversity is greatest in tropical latitudes. For example, an estimated 3,000 species of fish are found in the Amazon River alone.

Coral reef habitats also have extremely high biodiversity; nearly a quarter of all known marine species are found in coral reefs. The Great Barrier Reef, off the coast of Australia, is the largest coral reef system in the world. It supports over 700 species of coral, in addition to 1,600 fish species and 4,000 species of mollusks.

In the Antarctic Ocean, on the other hand, only 120 fish species are found. These species possess special molecular, biochemical "antifreeze" properties to deal with the cold water temperatures. However, Antarctic habitats nonetheless support many unique aquatic groups, such as the albatross, penguin, and large numbers of marine mammals such as the whale and seal.

Many fresh-water habitats also harbor a high proportion of unique species. This is due to the fact that, unlike oceans, fresh-water habitats often are isolated from one another, with natural barriers between them that are

The concept of biodiversity can be embodied in this colorful reef fish: a genetically distinct species is found in a specific habitat within a larger ecosystem. This particular fish could not survive, for example, in a polar sea or a tidal river. Difficult to cross. This results in the evolution of distinct species in different fresh-water habitats. The preservation of fresh-water habitats therefore is particularly critical to conserving aquatic biodiversity.

The Value of Biodiversity

The value of biodiversity is an issue that has caused considerable debate, given that the preservation of habitats often conflicts with the desires of developers. Yet there are several reasons for valuing biodiversity.

First, biodiversity is essential to the functioning of ecosystems. Each species plays a unique role within an ecosystem, and every species is dependent on others for food, shelter, or other resources. The loss of a single species therefore can have profound effects for the ecosystem as a whole.

Second, all species are potential sources of genetic variation for the development of new types of agricultural crops, as well as of medical drugs for treatment of human diseases. Third, biota (living organisms) have scientific and educational value. Finally, species have aesthetic and recreational value—consider, for example, the popularity of activities such as snorkeling, scuba diving, and hiking.

Scientists have shown that habitats with greater biodiversity are more resilient—that is, they are better able to adjust to and recover from various disturbances.

Because different species may perform overlapping functions in a biologically diverse ecosystem, a disturbance that affects one species may have lesser impact on the ecosystem as a whole. Habitats with little diversity are more vulnerable, because a disturbance affecting one species may cause the entire network of interactions to collapse.

Ecosystems approaches to natural resource management address interactions among species and among food webs, as well as the cycling of resources such as carbon, water, and nitrogen. These ecosystems approaches focus not on single species, but on the preservation of complex sets of interactions among species. Preservation of large, intact areas of habitat is necessary for the continued functioning of ecosystems.

Coral reefs are known for their biodiversity as well as their scientific, educational, and recreational value. This brain coral gets its name from its resemblance to the human brain.

Even a few centimeters of water can create a small habitat: in this case, a shallow coastal splash-zone pool on a rocky shore. Like tidal pools, such coastal environments support diverse communities of plants and animals.

Threats to Biodiversity

Aquatic biodiversity is threatened on many fronts. Fresh-water habitats support many of the most highly threatened animal groups. These include fish, mussels, amphibians, and crustaceans.

Fresh-water habitats are threatened by many factors, including pollution from industry, increased acidification, and agricultural runoff containing residues of fertilizers or pesticides. In addition, the building of dams destroys many river ecosystems. Development can harm aquatic habitats or remove them altogether, as when marshy areas are filled.

Aquatic ecosystems also are particularly fragile because the disturbance of a watershed can affect multiple components downstream, including rivers, lakes, estuaries, and oceans. Perhaps the largest threat to ocean biodiversity is overfishing.

In addition to depleting commercial species of fish, bivalves, and crustaceans, many fishing methods cause the needless deaths of noncommercial fish species as well as numerous reptiles, birds, and marine mammals.

WELLS AND WELL DRILLING

Approximately 97 per cent of the fresh water available in the world is underground. Wells provide groundwater for individual domestic needs, communities, cities, industry, crop irrigation, and agriculture. Some wells tap hot water, or geothermal resources. In other cases, groundwater is used solely for its cooling capabilities. Some wells are dug solely to study water quality or quantity: these are called monitoring wells or observation wells.

Regardless of its purpose, a well is defined as an artificial hole in the land surface created to access a liquid. It normally has a small diameter, typically less than 3 meters (10 feet), and usually measured in centimeters (inches). Wells may be constructed to seek water, oil, or natural gas.

Parts of a Well

The top portion of a well is commonly called the wellhead. The appearance of the wellhead varies depending on its purpose, when and how it was constructed, and what materials were available when it was built.

A hand-dug well may look like the nursery rhyme well in Jack and Jill—a deep hole surrounded by a stone wall. Most wells however, appear as a pipe, usually 5 to 25 centimeters (2 to 10 inches) in diameter, sticking up a short distance above the ground, typically less than 60 centimeters (2 feet).

In cold climates, the wellhead may be covered by a wellhouse or pumphouse to help minimize potential damage from freezing weather. When a well is constructed in an area where an above-ground structure would create a problem, such as in a road, the wellhead may be in a vault. The vault usually has a strong cement floor, with a drain, cement walls, and a steel top.

The typical components of a well include the sanitary seal, casing, casing seal, well screen, and the pump.

Casing, Screen, and Seals

The sanitary seal is the top covering of a well. Its function is to keep insects, small animals, dirt and debris from entering the well and contaminating the aquifer. This cap usually has an access port that allows periodic water-level measurements, and may have a screened breathing tube to allow air pressure to equalize as water levels rise and fall in the well.

The casing has several important functions. It is a conduit used to hold the wellhole open in soft or unconsolidated geologic formations. It commonly serves as the framework needed to support any pumping mechanism. It is an integral part of the casing seal, which is constructed to prevent surface water and shallow groundwater from entering the well and degrading deeper waters. To prevent surface runoff from flowing into the well, the casing usually extends 30 centimeters (1 foot) or more above the land surface. In formations prone to collapsing, such as in sands and gravel, the casing may extend the entire depth of the well. Where the well is constructed in hard formations such as shale or basalt, only the very top portion of the well may be cased.

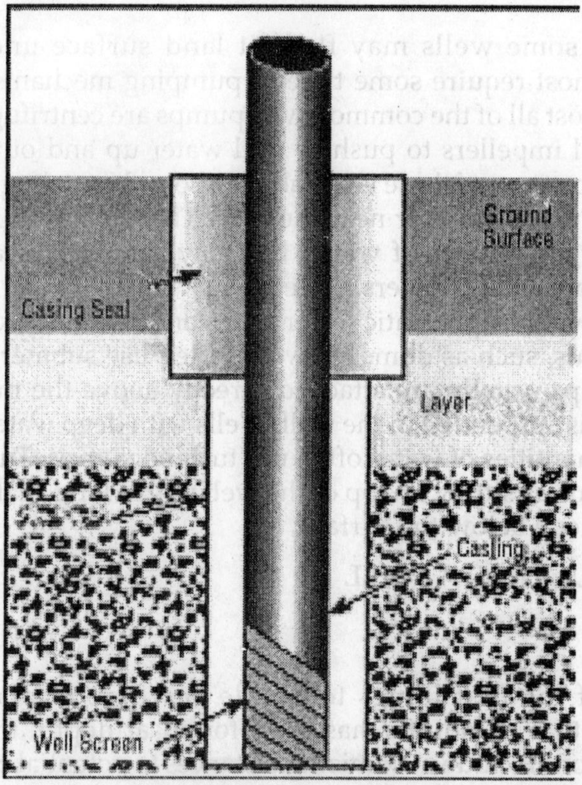

Even in ideal circumstances, a 5.5-meter-long (18-footlong) long casing generally is considered the minimum.

The casing seal is an impervious barrier, usually of cement grout or clay, placed between the well casing and the native material. It prevents surface water and undesirable groundwater from other aquifers from entering the wellbore or from flowing down the outside of the casing and degrading deeper water.

Where wells penetrate confining layers at greater depths, seals are commonly constructed at least 1.5 meters (5 feet) into a confining layer. Occasionally, seals are also constructed at greater depths between different aquifers. These seals prevent one aquifer from draining into another, potentially resulting in resource loss or degradation.

Screens in the casing are used to filter out sand and gravel while allowing maximum water flow into the well. Their construction varies from highly engineered screens over 15 meters (50 feet) long, to a few holes punched through the casing. The formations encountered in drilling and the intended use of the well indicate the type of screen, if any, that is needed.

Pumps

While some wells may flow at land surface under artesian pressure, most require some type of pumping mechanism to lift the water. Almost all of the common well pumps are centrifugal in design, using spiral impellers to push or pull water up and out of the well. The most common of these is the above-ground centrifugal, where the motor and pump sit at or near the well. These pumps are capable of moving large volumes of water, but they cannot typically lift from depths greater that 8 meters (25 feet).

Wells with deeper static water levels and relatively low discharge requirements, such as domestic wells, often use submersible pumps. These pumps usually are attached directly above the motor and the entire unit is suspended in the well. Wells with deep water levels used for large quantities of water often use turbine pumps. Turbine pumps have the motors sitting on top of the well with a long shaft turning the impellers deep below the surface.

CONSTRUCTING A WELL

Siting

One of the first things to decide is where to place the well. Assuming that an aquifer has been found at depth, the important considerations include maximizing distances to potential contaminants, access for machinery and potential future uses of the property. Septic systems, waste disposal lines, animal feedlots and buried fuel tanks all have the potential to degrade groundwater quality. Wells should always be at least 1.5 meters (5 feet) from any building site, in an area where future access can be preserved when repair, alteration or abandonment become necessary.

Construction

There are many ways to construct a satisfactory well. The intended use and local geology usually dictate which methods will work in a specific area. Some of the common construction methods include digging, driving, and drilling.

A dug well is built by excavating a hole into a water-bearing zone, lowering a casing into the aquifer, and constructing the seal. Dug wells usually

A worker reacts as water surges from a valve as a water well is activated in 1999 in Tekoa, West Bank. The well was the first to be owned and operated by the Palestinian Water Authority on the West Bank, and was financed by the U.S.

Agency for International Development. The water was earmarked for the Bethlehem and Hebron areas are inexpensive, but are restricted to the near-surface, which may be insufficient to provide ample water quantity of suitable quality.

A driven well typically consists of steel point attached to a section of perforated pipe or stiff screen, and sections of solid pipe. After digging or auguring an oversized hole to later accommodate the seal, the well constructor drives the assembly into the earth until the screen or perforations penetrate an aquifer; then he pours the seal. Most wells are constructed with well-drilling equipment. The two main types of drilling machines are the cable tool and the rotary drill.

The cable tool uses a heavy string of tools (bits), dropped repeatedly over the same point, to loosen, fracture and pulverize the earth so that it may then be lifted out of the hole with a bailing bucket. The loosened material and pulverized materials are referred to as drill cuttings. In dry formations, water may be added to the hole to hold the cuttings in suspension so they may be more easily lifted.

The bailing bucket is a long cylinder, fitted with a one-way valve at the bottom. The bucket is lowered into the hole, then lifted and drained as the hole is advanced.

Rotary drills use hardened drill bits to cut through the earth. Compressed air, water, or special water-based fluid called drilling mud is used to suspend the cuttings and lift them from the hole. This mud is usually forced down the inside of the drillstem, past the drillbit, where it picks up the cuttings, then up the outside of this drillstem to the land surface. The drilling mud also helps hold the hole open and forms a cake or barrier inside the hole so the fluid and cuttings are forced up the hole rather than into the formation.

Sealing

Regardless of which method is used, care must always be exercised in constructing the casing seal. Improperly sealed wells can contaminate water over a wide area. To construct the casing seal, the well constructor creates an oversized hole down to the intended sealing depth. The casing then is inserted, and the annulus (the space between the casing and earth) is filled with the sealing material.

Where the formation is prone to caving, another temporary casing may be necessary to hold the hole open while the seal is constructed. If the sealing depth is shallow (less than 18 meters [50 feet]) and there is little standing water in the sealing area, a special type of clay, known as Bentonite, may be used as the sealing material.

Dry Bentonite chips or pellets are slowly poured into the annular space between the casing and the natural formation. The other common sealing material is cement grout, which is similar to concrete without the gravel component. This fine-grained cement is pumped into the bottom of the annulus until it reaches the land surface.

Developing

Once the well is drilled and the physical components have been constructed, the well must be developed. The purpose of developing the well is to maximize its discharge capacity. This usually is done by alternately pumping the well and surging water in the bore to wash silt and sand from the formation and open it to greater flow rates. A properly developed well will produce adequate quantities of sand and mud-free water.

Testing

After completion of construction, a well test is performed to determine the capacity of the well to yield water over time. This can be done with a pump, a bailer bucket, or by blowing compressed air into the well bore. A component of this test is the monitoring of the well's drawdown.

Drawdown is the change in water level in the well due to pumping. A 1-hour test is considered the minimum period necessary to evaluate a well, with 4-hour and 24-hour tests being more reliable. This information is necessary to determine the usefulness of the well for its desired purpose and the size of pump it can support.

Disindfecting

The well construction process will invariably introduce bacteria into the wellbore. Once construction is completed, the casing, wellbore, and the standing water in the well are sterilized, usually with chlorine or hypochlorate solution.

Once the disinfectant has had sufficient opportunity to sterilize the well, the well is purged, and the fresh water usually is tested for biological and chemical purity.

Logging

During construction, development and testing of the well, it is very important that the constructor record every detail about formations encountered, water-bearing zones, construction components, and well yield.

This record is called the water-well report or well log. It is extremely useful if the well ever needs repair or is to be abandoned. It also provides valuable information to the landowner, other well drillers, and resource managers.

Repair and Abandonment

Over time, old and improperly constructed wells will deteriorate. The landowner is responsible for any wells on his or her property. Wells with faulty seals or that are commingling aquifers may deplete or contaminate the resource.

Depending on the original construction methods and materials used, a defective well may be repaired by a skilled well constructor. In many cases, it is easier and cheaper to permanently abandon a defective well than to repair it.

Permanent abandonment typically involves removing the existing casing, cleaning out the hole, and completely filling it with cement grout. In situations where the casing can not be removed, the casing must be perforated so the cement grout can penetrate into the formations and prevent any vertical movement of water in the wellbore.

Drilling our Own Well

A water well is much more than a hole in the ground. To prevent groundwater contamination, a well constructor must use proper methods and equipment. Licensed and bonded water-well constructors have the equipment, knowledge, and experience required for proper well construction.

For this reason, most states discourage landowners from drilling a well themselves. If a landowner decides to drill, alter, or abandon a well on his own property, he should first consult with the local regulating authorities to become familiar with specific construction requirement for the area.

WASTEWATER TREATMENT AND MANAGEMENT

Waters that are used for drinking, manufacturing, farming, and other purposes are degraded in quality as a result of the introduction of contaminating constituents. Organic wastes, suspended solids, bacteria, nitrates, and phosphates are pollutants that commonly must be removed.

To make wastewater acceptable for reuse or for returning to the environment, the concentration of contaminants must be reduced to a nonharmful level, usually a standard prescribed by the U.S. Environmental Protection Agency. Furthermore, urban stormwaters flowing over lawns, rooftops, and paved surfaces are polluted by lawn chemicals, oil and gasoline spills on streets, plus other substances that become entrained in them as they make their way to a stream, river, or lake.

These flows must also be subjected to some form of treatment to make them less harmful to the environment. Restoration of water quality is accomplished through the use of a variety of pollution control methods.

In urbanized areas, municipal wastewaters (mainly sewage) generally are conveyed to a point of treatment through sanitary sewers, whereas stormwaters are conveyed to their receiving bodies of water through storm drainage networks.

In the past, cities sometimes used combined wastewater collection systems wherein a single sewerage network collected domestic wastewater, industrial wastes, and storm runoff water. But this configuration does not support the level of pollution control required today, and new systems of this type are no longer being built.

Sanitary sewers carry some level of flow during all hours of the day and night, whereas storm sewers flow mainly after periods of rainfall. During major storm events, the volumes of water carried by storm sewers are orders of magnitude greater than those carried by sanitary sewers. Wastewaters and stormwaters are subjected to treatment, but the types of treatment generally are quite different.

Wastewater Treatment Process

The task of designing and constructing facilities for treating wastewaters falls to environmental engineers. They employ a variety of engineered and natural systems to get the job done, using physical, chemical, biological, and sludge treatment methods.

The features of wastewater treatment systems are determined by:

(1) The nature of the municipal and industrial wastes that are conveyed to them by sewers,

(2) The amount of treatment required to preserve and/or improve the quality of the receiving bodies of water.

Discharges from treatment plants usually are disposed by dilution in rivers, lakes, or estuaries. They also may

Pumps are used to aerate wastewater in a treatment process unit. Aeration supports biological and other treatment processes be used for certain types of irrigation (such as golf courses), transported to lagoons where they are evaporated, or discharged through submarine (underwater) outfalls into the ocean.

However, outflows from treatment works must meet effluent standards set by the U.S. Environmental Protection Agency to avoid polluting the bodies of water that receive them.

The categories of wastewater treatment are primary, secondary, and tertiary, or advanced. The minimum level of treatment required is usually secondary treatment, but some cities and industries are required to install tertiary or advanced wastewater treatment processes for removal of pollutants that are resistant to conventional treatment.

Stream classification documents, published by each state as required by the U.S. Clean Water Act of 1977, categorize surface waters according to their most beneficial present or future use, such as for drinking-water supplies, body-contact recreation, and so on.

These publications also incorporate stream standards that establish maximum allowable pollutant concentrations for a given stream under defined flow conditions. Effluent standards under the Clean Water Act's National Pollutant Discharge Elimination System (NPDES) are used for regulatory purposes to achieve compliance with these stream standards. NPDES permits are issued to cities or other facilities that regulate the volume of discharge, contaminant concentrations, and timing of discharge so as to protect water quality in the receiving waterbody.

Conventional Treatment

Conventional wastewater treatment consists of preliminary processes, primary settling to remove heavy solids and floatable materials, and secondary biological aeration to metabolize and flocculate colloidal and dissolved organics. Waste sludge drawn from these operations is thickened and processed for ultimate disposal, usually either land application or landfilling.

Preliminary treatment processes include coarse screening, medium screening, shredding of solids, flow measuring, pumping, grit removal, and preaeration. Chlorination of raw wastewater sometimes is used for odor control and to improve settling characteristics of the solids.

An aerial view depicts a typical wastewater treatment plant supporting primary and secondary treatment. Visible in this photograph are flocculation basins (rectangular), primary settling tanks (foreground), and secondary trickling filter).

Primary and Secondary Treatment

Primary treatment involves sedimentation, and is the process by which about 30 to 50 per cent of the suspended solid materials in raw wastewater are removed. Sedimentation must precede all biological filtration operations.

The organic matter remaining after primary treatment is extracted by biological secondary treatment processes to meet effluent standards.

Secondary treatment commonly is carried out using activated-sludge processes, trickling filters, or rotating biological contactors.

In the activated-sludge method, wastewater is fed continuously into an aerated tank where microorganisms break down the organics. The resulting microbial floc (activated sludge) is settled under quiescent (calm-water) conditions in a final clarifier and returned to an aeration tank. The plant effluent is clear supernatant from secondary settling.

Trickling filters and rotating biological contactors have media to support microbial films. These slime growths extract organic materials from wastewater as it trickles over the surfaces. Oxygen is supplied from air moving through voids (empty spaces) in the media. Excessive biological growth washes out and is collected in a secondary clarifier.

Tertiary Treatment

Tertiary wastewater treatment is additional treatment that follows primary and secondary treatment processes. It is employed when primary and secondary treatment cannot accomplish all that is required. For example, phosphorus removal may be needed for wastewaters that are discharged to receiving waters that are likely to become eutrophic, or enriched with nutrients.

(Cultural or human-enhanced eutrophication often is associated with nitrogen and phosphorous in effluent.) Water reclamation is achieved in varying degrees, but only a few large-scale plants are reclaiming water to near-pristine quality.

Sludge Processing and Disposal

Primary sedimentation and secondary biological flocculation processes concentrate waste organics into a volume of sludge significantly less than the quantity of wastewater treated. But disposal of the accumulated waste sludge is a major economic factor in wastewater treatment.

Methods for processing raw sludge include anaerobic (biological) digestion and mechanical dewatering by either belt-filter pressing or centrifugation. Conventional methods of disposal are application as a fertilizer or soil conditioner on agricultural land, landfilling in a dedicated disposal site, or codisposal with municipal solid waste.

Stormwater Treatment and Management

Stormwater treatment includes:

- Storage in retention ponds where evaporation and seepage take place.
- Diversion to natural or artificial wetlands, where pollutants

are removed by vegetation and sedimentation and water is returned to the atmosphere by evapotranspiration.

These methods take advantage of the ability of natural filtration and biological processes to aid in restoring water quality. Under certain circumstances, chemicals may also be introduced as treatment aids.

The principal method used for stormwater treatment is storage wherein natural processes of sedimentation, evaporation, and nutrient removal take place. Because of the large volumes of water generated by storms, it usually is not practical to divert these waters to treatment plants such as those used to process municipal and industrial wastewaters.

However, a number of devices can be inserted into stormwater systems to achieve various levels of removal of solids and other constitutents. These devices employ features of some of the components of wastewater treatment plants described previously.

LANDFILLS: IMPACT ON GROUNDWATER

Solid waste landfills are a necessity in modern-day society, because the collection and disposal of waste materials into centralized locations helps minimize risks to public health and safety. Solid waste landfills, which are regulated differently than hazardous waste landfills, may accept a variety of solid, semi-solid, and small quantities of liquid wastes.

Landfills generally remain open for decades before undergoing closure and postclosure phases, during which steps are taken to minimize the risk of environmental contamination.

Municipal solid waste (MSW) landfills accept nonhazardous wastes from a variety of sources, such as households, businesses, restaurants, medical facilities, and schools. Many MSW landfills also can accept contaminated soil from gasoline spills, conditionally exempted hazardous waste from businesses, small quantities of hazardous waste from households, and other toxic wastes.

Industrial facilities may utilize their own captive landfill (i.e., a solid waste landfill for their exclusive use) to dispose of nonhazardous waste from their processes, such as sludge from paper mills and wood waste from woodprocessing facilities.

The Concern Over Landfill Impacts

Although landfills are an indispensable part of everyday living, they may present long-term threats to groundwater and also surface waters that are hydrologically connected. In the United States, federal standards to protect groundwater quality were implemented in 1991 and required some landfills to use plastic liners and collect and treat leachate. However, many disposal sites were either exempted from these rules or grandfathered (excused from the rules owing to previous usage).

Although the federal rules marked a significant improvement in the management of solid waste, some think that these rules do not go far enough. There is an increasing belief among solid waste experts that unless further steps are taken to detoxify landfilled materials, today's society will be placing a burden on upcoming generations to address future landfill impacts. Much of the concern revolves around leachate, the watery solution that results after water passes through a landfill.

Leachate Generation and Composition

The precipitation that falls into a landfill, coupled with any disposed liquid waste, results in the extraction of the water-soluble compounds and particulate matter of the waste, and the subsequent formation of leachate. The creation of leachate, sometimes deemed "garbage soup," presents a major threat to the current and future quality of groundwater.

Parameter	Typical Range (milligrams per liter, unless otherwise noted)	Upper Limit (milligrams per liter, unless otherwise noted)
Total Alkalinity (as $CaCO_3$)	730–15,050	20,850
Calcium	240–2,330	4,080
Chloride	47–2,400	11,375
Magnesium	4–780	1,400
Sodium	85–3,800	7,700
Sulfate	20–730	1,826
Specific Conductance	2,000–8,000 μmhos/cm	9,000 μmhos/cm
Total Dissolved Solids	1,000–20,000	55,000
Chemical Oxygen Demand	100–51,000	99,000
Biological Oxygen Demand	1,000–30,300	195,000
Iron	0.1–1,700	5,500
Total Nitrogen	2.6–945	1,416
Potassium	28–1,700	3,770
Chromium	0.5–1.0	5.6
Manganese	Not detected – 400	1,400
Copper	0.1–9.0	9.9
Lead	Not detected – 1.0	14.2
Nickel	0.1–1.0	7.5

(Other major threats include underground storage tanks,

abandoned hazardous waste sites, agricultural activities, and septic tanks.)

Leachate composition varies relative to the amount of precipitation and the quantity and type of wastes disposed. In addition to numerous hazardous constituents, leachate generally contains nonhazardous parameters that are also found in most groundwater systems. These constituents include dissolved metals (e.g., iron and manganese), salts (e.g., sodium and chloride), and an abundance of common anions and cations (e.g., bicarbonate and sulfate).

However, these constituents in leachate typically are found at concentrations that may be an order of magnitude (or more) greater than concentrations present in natural groundwater systems.

Leachate from MSW landfills typically has high values for total dissolved solids and chemical oxygen demand, and a slightly low to moderately low pH.

MSW leachate contains hazardous constituents, such as volatile organic compounds and heavy metals. Wood-waste leachates typically are high in iron, manganese, and tannins and lignins. Leachate from ash landfills is likely to have elevated pH and to contain more salts and metals than other leachates.

Leachate Release and Migration

A release of leachate to the groundwater may present several risks to human health and the environment. The release of hazardous and nonhazardous components of leachate may render an aquifer unusable for drinking-water purposes and other uses. Leachate impacts to groundwater may also present a danger to the environment and to aquatic species if the leachate-contaminated groundwater plume discharges to wetlands or streams.

Once leachate is formed and is released to the groundwater environment, it will migrate downward through the unsaturated zone until it eventually reaches the saturated zone. Leachate then will follow the hydraulic gradient of the groundwater system.

Monitoring wells at landfills allow scientists to determine whether contaminants in leachate are escaping into the local groundwater system. The wells are placed downgradient of the landfill at appropriate depths and at various intervals to intercept any contaminants and monitor their movement.

A number of forces may act on or react with the migrating leachate, resulting in changes of chemistry and a general reduction of strength from the original release. These forces are physical (filtration, sorption, advection, and dispersion), chemical (oxidation-reduction, precipitation-dissolution, adsorption-desorption, hydrolysis, and ion exchange), and biological (microbial degradation).

The extent of these reactions depends on the materials underlying the landfill, the hydraulics of the groundwater system, and the chemistry of the leachate.

Although many of these reactions have the capability to reduce the potential impact to groundwater, some (such as microbial degradation) can actually increase the toxicity by producing by-products that are more hazardous than the original contaminant. This can be seen, for example, in the creation of vinyl chloride from the degradation of trichloroethene.

Old and New Viewpoints

Today's landfills are constructed with liners that contain leachate, and leachate collection systems that collect it. But historically, many landfills were constructed without liners or leachate collection systems.

Two philosophies previously existed regarding the placement of unlined landfills. One viewpoint was to place most of these facilities in moderately permeable materials and as close as possible to rivers or streams (i.e., surface water). This type of siting would allow the landfills to slowly (and deliberately) leak leachate into the groundwater, minimizing the length and size of the leachate plume, which would ultimately discharge into surface water.

The other theory was the exact opposite: to place landfills as far away from surface water as possible. This may have spared impacts to the surface water, but locating unlined facilities in this fashion usually resulted in creating significant leachate-contaminated groundwater plumes that followed the direction of groundwater flow.

Are Design Standards Enough ?

The U.S. Environmental Protection Agency issued standards for MSW landfill design and operation in 1991. These new rules were adopted to protect groundwater from the release of leachate by MSW landfills. Landfill owners who could not meet these new design standards requiring plastic liners and leachate collection were required to close and to conduct groundwater monitoring for 30 years

. Groundwater protection standards were developed for all MSW landfills, setting a national precedent for solid waste management that may eventually become the standard for all types of landfills.

As the twenty-first century opened, there was a controversy whether the new rules for MSW landfills offer adequate environmental protection, especially as these facilities age.

Landfill designs that utilize plastic liners below the waste and then are covered with plastic when the landfill stops placing waste into the active cell (area of waste input) are referred to as "dry tombs." These engineered systems are designed to minimize leachate generation by restricting the introduction of moisture, primarily precipitation.

It is widely recognized that even the best-installed plastic liner will succumb to deterioration and eventually will allow leachate to be created and released.

However, this may not happen within the required 30 years of postclosure groundwater monitoring. Moreover, it may not be detected during the time the landfill operators are actively involved and financially obligated.

Opponents of dry-tomb landfills advocate for recycling the collected leachate through the waste, which will enhance the rate of chemical reactions inside the landfill and eventually stabilize the waste material prior to covering the landfill. This could reduce the toxic nature of the waste materials in the closed landfill and minimize the future threat posed to groundwater from these facilities.

RUNOFF, FACTORS AFFECTING

Rain falling on the landscape may flow quickly over soil or rock surfaces as runoff to stream channels. Alternately, some water may flow more slowly downslope toward streams within the soil.

Some may percolate downward through pores in soil and fractures in rock to reach the top of the saturated zone (often called the water table). Below the saturated zone, it flows much more slowly as groundwater.

Factors Influencing Infiltration Rate

Soil characteristics, plants and animals, and slope angle are among the natural factors controlling the proportion of precipitation that is converted to runoff in a given landscape, and the time it takes for runoff to enter a stream. Human changes to these landscape features can greatly influence runoff. Grassed filter strips in farm fields help reduce runoff and erosion by slowing water velocities in the vegetated areas. Grassy strips also reduce erosion by trapping excess sediment, nutrients, and farm chemicals. Vegetation (including dead vegetative materials) in prairies, forests, and other natural areas plays a similar role.

The soil surface acts as a filter that lets water pass through (infiltrate) at a rate known as the infiltration rate or infiltration capacity. Runoff may be produced when precipitation or snowmelt adds water to the soil surface faster than it can be absorbed.

The excess water remains on the surface and flows downslope as runoff. For example, if the precipitation rate is 5 centimeters (about 2 inches) per hour, but the infiltration rate is only 2.5 centimeters (about 1 inch) per hour, surface runoff is produced at the rate of 2.5 centimeters (about 1 inch) per hour, even if the soil is not entirely saturated. This mechanism of runoff generation is more common in drier climates where vegetation cover is sparse. In humid areas with greater vegetation cover, the water table may lie at the surface in low-lying areas or slope hollows, so that the soil there is saturated. Saturated areas expand during rain or snowmelt, as well as during the cold season when plants withdraw little water from the soil.

Any rain that falls on these saturated areas must run off over the surface. In times of prolonged heavy rainfall, large areas of a gently sloping landscape may become saturated, and much of the rain that follows runs off rapidly to streams. This was the case during the devastating Mississippi River flood of 1993, when much of the landscape in the Upper Mississippi River Basin appeared as a "lake" on satellite images that detect surface water.

Soil Characteristics

Infiltration rate is controlled by the nature of the soil, by the plant and animal communities it supports, and by human influences. Where soil is absent and little-fractured bedrock is exposed, water cannot soak in and will run off rapidly. If soil is present, but is very fine-grained and clay-rich, the pore spaces that water must pass through are extremely small; hence, water will infiltrate very slowly compared to sandy soils that readily soak up water.

Some finer-grained soils have vertical cracks that form when the soil shrinks as it dries. These cracks allow water to enter more readily, but may close up after the soil is wetted.

Compaction of soils reduces the size of pore spaces and the infiltration rate. Water commonly runs off areas that were compacted through repeated passage of people, large animals, or heavy machinery.

Raindrops falling on bare soil also can compact the soil surface in plowed fields, leading to increased runoff and erosion of farmland.

Ice and Snow

Ice within the pore spaces of frozen soils typically reduces infiltration rates, but only soils that were saturated on freezing will completely prevent water from entering. Summer melting of the upper layer of permafrost in cold regions often results in a saturated zone of soil above the frozen ground.

Precipitation that falls as snow is stored until snowmelt, when a large pulse of runoff may be generated. Runoff occurs only after the entire snowpack has reached 0°C (32°F), some melting has occurred, and pore spaces between snow grains can no longer hold all the water supplied.

The rate of snowmelt depends on heat inputs into the snowpack through solar radiation and via water from melting and rainfall moving through the snow. The soil surface underneath the melting snowpack may become saturated, such that runoff flows through the base of the snowpack toward streams. Saturated zones and surface runoff commonly occur on slopes below the melting snow line.

Plants and Animals

In general, plants and small animals tend to increase the infiltration rate of soils. Some water usually evaporates from plant surfaces before it can fall to the soil surface. A plant cover and litter layer of dead vegetation protects the soil surface from compaction by heavy raindrops, and also slows the delivery of water to the soil surface. Plant stems help slow down water that flows over the soil surface.

Plant roots help create openings in the soil, and also draw water from beneath the soil surface and transpire it through leaves back to the atmosphere. Decayed plant matter helps keep fine soil particles (such as clay) from sticking together, thereby increasing infiltration capacity. The burrowing activities of small animals such as insects, worms, and gophers also help keep the soil loose and create small openings through which water can pass.

When the landscape is completely devegetated, for example, following a forest fire or during a construction project, a dramatic increase in runoff and soil erosion may result. In desert environments where much of the soil surface lacks vegetation and where bare rock is exposed, most of the rainfall in heavy thunderstorms runs off rapidly and flash floods are common. Yet in dense, humid forests, vegetation and thick, loose soils may absorb water so readily that water rarely runs off the surface.

Slope

Steep slopes in the headwaters of drainage basins tend to generate more runoff than do lowland areas. Mountain areas tend to receive more precipitation overall because they force air to be lifted and cooled. On gentle slopes, water may temporarily pond and later soak in. But on steep mountainsides, water tends to move downward more rapidly.

Soils tend to be thinner on steep slopes, limiting storage of water, and where bedrock is exposed, little infiltration can occur. In some cases, however, accumulations of coarse sediment at the base of steep slopes soak up runoff from the cliffs above, turning it into subsurface flow.

RUNOFF AND FLOODING

Water commonly flows downslope through the loose soil overlying bedrock.

Rain falling on areas where unfractured bedrock is exposed has little opportunity to infiltrate, and instead will run off the surface. A brief thunderstorm in Yellowstone National Park produced considerable surface runoff from these steep cliffs much less likely to cause flooding, but is faster than the creeping flow of groundwater in the bedrock below.

RUNOFF AND URBAN DEVELOPMENT

Urban development can greatly increase the amount of precipitation that is converted to runoff in a drainage basin. Most paved surfaces and rooftops allow no water to infiltrate, but instead divert water directly to storm channels and drains. Urbanization is of serious concern to water resources for several reasons.

- First, the increased amount of water flowing to streams during storms causes larger floods, and floods build to a peak faster because of the rapid flow of water over smooth surfaces.
- Second, motor vehicles leave oils and exhaust residues on streets, and household and industrial chemicals also collect on pavement surfaces. These nonpoint-source pollutants are readily washed off during storms, contaminating streams into which urban runoff flows. Careless disposal of hazardous wastes on streets or in storm drains adds to the problem.
- Third, most precipitation has no chance to percolate downward to groundwater, so the supply of groundwater to wells is reduced.

Some cities have taken steps to reduce these impacts. Pavement can be constructed so that some water passes through to recharge groundwater, and storm runoff can be routed to artificial basins that allow water to soak in. Along with regulation of hazardous industrial wastes, programs have been developed to educate the public about the dangers of improper disposal of wastes on streets and in storm drains.

Chapter 3

Microbial Ecology

Microbial ecology is the relationship of microorganisms with one another and with their environment. It concerns the three major domains of life — Eukaryota, Archaea, and Bacteria — as well as viruses. Microorganisms, by their omnipresence, impact the entire biosphere. They are present in virtually all of our planet's environments, including some of the most extreme, from acidic lakes to the deepest ocean, and from frozen environments to hydrothermal vents.

Microbes, especially bacteria, often engage in symbiotic relationships (either positive or negative) with other organisms, and these relationships affect the ecosystem. One example of these fundamental symbioses are chloroplasts, which allow eukaryotes to conduct photosynthesis. Chloroplasts are considered to be endosymbiotic cyanobacteria, a group of bacteria that are thought to be the origins of aerobic photosynthesis.

Some theories state that this invention coincides with a major shift in the early earth's atmosphere, from a reducing atmosphere to an oxygen-rich atmosphere. Some theories go as far as saying that this shift in the balance of gasses might have triggered a global ice-age known as the Snowball Earth.

They are the backbone of all ecosystems, but even more so in the zones where light cannot approach and thus photosynthesis cannot be the basic means to collect energy. In such zones, chemosynthetic microbes provide energy and carbon to the other organisms.

Other microbes are decomposers, with the ability to recycle nutrients from other organisms' waste poducts. These microbes play a vital role in biogeochemical cycles. The nitrogen cycle, the phosphorus cycle and the carbon cycle all depend on microorganisms in one way or another. For example, nitrogen which makes up 78% of the planet's atmosphere is "indigestible" for most organisms, and the flow of nitrogen into the biosphere depends on a microbial process called fixation.

Due to the high level of horizontal gene transfer among microbial communities, microbial ecology is also of importance to studies of evolution.

MICROBIAL RESOURCE MANAGEMENT

Biotechnology may be used alongside microbial ecology to address a number of environmental and economic challenges. Managing the carbon cycle to sequester carbon dioxide and prevent excess methanogenesis is important in mitigating global warming, and the prospects of bioenergy are being expanded by the development of microbial fuel cells.

Microbial resource management advocates a more progressive attitude towards disease, whereby biological control agents are favoured over attempts at eradication. Fluxes in microbial communities has to be better characterized for this field's potential to be realised.

MICROBIAL BIODEGRADATION

Interest in the microbial biodegradation of pollutants has intensified in recent years as humanity strives to find sustainable ways to cleanup contaminated environments. These bioremediation and biotransformation methods endeavour to harness the astonishing, naturally occurring, ability of microbial xenobiotic metabolism to degrade, transform or accumulate a huge range of compounds including hydrocarbons (e.g. oil), polychlorinated biphenyls (PCBs), polyaromatic hydrocarbons (PAHs), pharmaceutical substances, radionuclides and metals.

Major methodological breakthroughs in recent years have enabled detailed genomic, metagenomic, proteomic, bioinformatic and other high-throughput analyses of environmentally relevant microorganisms providing unprecedented insights into key biodegradative pathways and the ability of organisms to adapt to changing environmental conditions.

The elimination of a wide range of pollutants and wastes from the environment is an absolute requirement to promote a sustainable development of our society with low environmental impact. Biological processes play a major role in the removal of contaminants and they take advantage of the astonishing catabolic versatility of microorganisms to degrade/convert such compounds.

New methodological breakthroughs in sequencing, genomics, proteomics, bioinformatics and imaging are producing vast amounts of information. In the field of Environmental Microbiology, genome-based global studies open a new era providing unprecedented *in silico* views of metabolic and regulatory networks, as well as clues to the evolution of degradation pathways and to the molecular adaptation strategies to changing environmental conditions.

Functional genomic and metagenomic approaches are increasing our understanding of the relative importance of different pathways and regulatory networks to carbon flux in particular environments and for particular compounds and they will certainly accelerate the development of bioremediation technologies and biotransformation processes.

Aerobic Biodegradation of Pollutants

The burgeoning amount of bacterial genomic data provides unparalleled opportunities for understanding the genetic and molecular bases of the degradation of organic pollutants. Aromatic compounds are among the most recalcitrant of these pollutants and lessons can be learned from the recent genomic studies of *Burkholderia xenovorans* LB400 and *Rhodococcus* sp. strain RHA1, two of the largest bacterial genomes completely sequenced to date.

These studies have helped expand our understanding of bacterial catabolism, non-catabolic physiological adaptation to organic compounds, and the evolution of large bacterial genomes. First, the metabolic pathways from phylogenetically diverse isolates are very similar with respect to overall organization.

Thus, as originally noted in pseudomonads, a large number of "peripheral aromatic" pathways funnel a range of natural and xenobiotic compounds into a restricted number of "central aromatic" pathways. Nevertheless, these pathways are genetically organized in genus-specific fashions, as exemplified by the b-ketoadipate and Paa pathways.

Comparative genomic studies further reveal that some pathways are more widespread than initially thought. Thus, the Box and Paa pathways illustrate the prevalence of non-oxygenolytic ring-cleavage strategies in aerobic aromatic degradation processes. Functional genomic studies have been useful in establishing that even organisms harboring high numbers of homologous enzymes seem to contain few examples of true redundancy.

For example, the multiplicity of ring-cleaving dioxygenases in certain rhodococcal isolates may be attributed to the cryptic aromatic catabolism of different terpenoids and steroids. Finally, analyses have indicated that recent genetic flux appears to have played a more significant role in the evolution of some large genomes, such as LB400's, than others. However, the emerging trend is that the large gene repertoires of potent pollutant degraders such as LB400 and RHA1 have evolved principally through more ancient processes. That this is true in such phylogenetically diverse species is remarkable and further suggests the ancient origin of this catabolic capacity.

Anaerobic Biodegradation of Pollutants

Anaerobic microbial mineralization of recalcitrant organic pollutants is of great environmental significance and involves intriguing novel biochemical reactions. In particular, hydrocarbons and halogenated compounds have long been doubted to be degradable in the absence of oxygen, but the isolation of hitherto unknown anaerobic hydrocarbon-degrading and reductively dehalogenating bacteria during the last decades provided ultimate proof for these processes in nature.

Many novel biochemical reactions were discovered enabling the respective metabolic pathways, but progress in the molecular understanding of these bacteria was rather slow, since genetic systems are not readily applicable for most of them. However, with the increasing application of genomics in the field of environmental microbiology, a new and promising perspective is now at hand to obtain molecular insights into these new metabolic properties.

Several complete genome sequences were determined during the last few years from bacteria capable of anaerobic organic pollutant degradation. The ~4.7 Mb genome of the facultative denitrifying *Aromatoleum aromaticum* strain EbN1 was the first to be determined for an anaerobic hydrocarbon degrader (using toluene or ethylbenzene as substrates).

The genome sequence revealed about two dozen gene clusters (including several paralogs) coding for a complex catabolic network for anaerobic and aerobic degradation of aromatic compounds. The genome sequence forms the basis for current detailed studies on regulation of pathways and enzyme structures. Further genomes of anaerobic hydrocarbon degrading bacteria were recently completed for the iron-reducing species *Geobacter metallireducens* and the perchlorate-reducing *Dechloromonas aromatica*, but these are not yet evaluated in formal publications.

Complete genomes were also determined for bacteria capable of anaerobic degradation of halogenated hydrocarbons by halorespiration: the ~1.4 Mb genomes of *Dehalococcoides ethenogenes* strain 195 and *Dehalococcoides* sp. strain CBDB1 and the ~5.7 Mb genome of *Desulfitobacterium hafniense* strain Y51. Characteristic for all these bacteria is the presence of multiple paralogous genes for reductive dehalogenases, implicating a wider dehalogenating spectrum of the organisms than previously known.

Moreover, genome sequences provided unprecedented insights into the evolution of reductive dehalogenation and differing strategies for niche adaptation.

Bioavailability, Chemotaxis, and Transport of Pollutants

Bioavailability, or the amount of a substance that is physiochemically accessible to microorganisms is a key factor in the efficient biodegradation of pollutants. Chemotaxis, or the directed movement of motile organisms towards or away from chemicals in the environment is an important physiological response that may contribute to effective catabolism of molecules in the environment. In addition, mechanisms for the intracellular accumulation of aromatic molecules via various transport mechanisms are also important.

Oil Biodegradation

Petroleum oil contains aromatic compounds that are toxic for most life forms. Episodic and chronic pollution of the environment by oil causes major ecological perturbations. Marine environments are especially vulnerable since oil spills of coastal regions and the open sea are poorly containable and mitigation is difficult. In addition to pollution through human activities, millions of tons of petroleum enter the marine environment every year from natural seepages.

Despite its toxicity, a considerable fraction of petroleum oil entering marine systems is eliminated by the hydrocarbon-degrading activities of microbial communities, in particular by a remarkable recently discovered group of specialists, the so-called hydrocarbonoclastic bacteria (HCB). *Alcanivorax borkumensis* was the first HCB to have its genome sequenced.

Analysis of Waste Biotreatment

Sustainable development requires the promotion of environmental management and a constant search for new technologies to treat vast quantities of wastes generated by increasing anthropogenic activities. Biotreatment, the processing of wastes using living organisms, is an environmentally friendly, relatively simple and cost-effective alternative to physico-chemical clean-up options.

Confined environments, such as bioreactors], have been engineered to overcome the physical, chemical and biological limiting factors of biotreatment processes in highly controlled systems. The great versatility in the design of confined environments allows the treatment of a wide range of wastes under optimized conditions. To perform a correct assessment, it is necessary to consider various microorganisms having a variety of genomes and expressed transcripts and proteins.

A great number of analyses are often required. Using traditional genomic techniques, such assessments are limited and time-consuming. However, several high-throughput techniques originally developed for medical studies can be applied to assess biotreatment in confined environments.

Metabolic Engineering and Biocatalytic Applications

The study of the fate of persistent organic chemicals in the environment has revealed a large reservoir of enzymatic reactions with a large potential in preparative organic synthesis, which has already been exploited for a number of oxygenases on pilot and even on industrial scale. Novel catalysts can be obtained from metagenomic libraries and DNA sequence based approaches.

Our increasing capabilities in adapting the catalysts to specific reactions and process requirements by rational and random mutagenesis broadens the scope for application in the fine chemical industry, but also in the field of biodegradation. In many cases, these catalysts need to be exploited in whole cell bioconversions or in fermentations, calling for system-wide approaches to understanding strain physiology and metabolism and rational approaches to the engineering of whole cells as they are increasingly put forward in the area of systems biotechnology and synthetic biology.

CHEMOTAXIS

Chemotaxis, a kind of taxis, is the phenomenon in which bodily cells, bacteria, and other single-cell or multicellular organisms direct their movements according to certain chemicals in their environment. This is important for bacteria to find food (for example, glucose) by swimming towards the highest concentration of food molecules, or to flee from poisons (for example, phenol). In multicellular organisms, chemotaxis is critical to early (e.g. movement of sperm towards the egg during fertilization) and subsequent phases of development (e.g. migration of neurons or lymphocytes) as well as in normal function. In addition, it has been recognized that mechanisms that allow chemotaxis in animals can be subverted during cancer metastasis.

Chemotaxis is called positive if movement is in the direction of a higher concentration of the chemical in question, and negative if the direction is opposite.

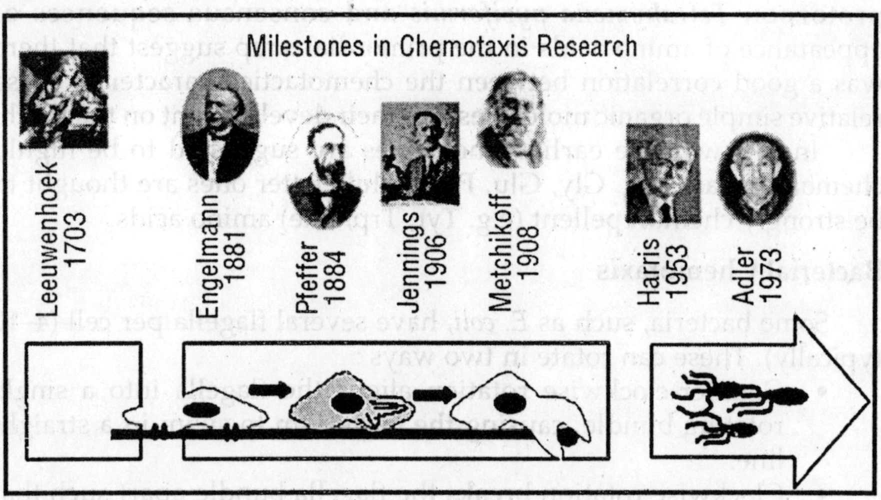

Milestones of Chemotaxis Research

Although migration of cells was detected from the early days of the development of microscopy *(Leeuwenhoek)*, erudite description of chemotaxis was first made by *T.W. Engelmann* (1881) and *W.F. Pfeffer* (1884) in bacteria and *H.S. Jennings* (1906) in ciliates. The Nobel Prize Laureate *E. Metchnikoff* also contributed to the study of the field with investigations of the process as an initial step of phagocytosis.

The significance of chemotaxis in biology and clinical pathology was widely accepted in the 1930s. The most fundamental definitions belonging to the phenomenon were also drafted by this time. The most important aspects in quality control of chemotaxis assays were described by *H. Harris* in the 1950s. In the 1960s and 1970s, the revolution of modern cell biology and biochemistry provided a series of novel techniques which became available to investigate the migratory responder cells and subcellular fractions responsible for chemotactic activity.

The pioneering works of *J. Adler* represented a significant turning point in understanding the whole process of intracellular signal transduction of bacteria.

Phylogeny and Chemotactic Signaling

Chemotaxis is one of the most basic cell physiological responses. Development of receptor systems for the detection of harmful and favorable substances in the environment was most essential to unicellular organisms from the very early stages of phylogeny. Comprehensive analysis of chemotactic activity of the eukaryotic protozoon *Tetrahymena pyriformis* and consensus sequences of appearance of amino acids in the primordial soup suggest that there was a good correlation between the chemotactic character of these relative simple organic molecules and their development on the Earth.

In this way the earliest molecules are suggested to be highly chemoattractant (e.g. Gly, Glu, Pro), while latter ones are thought to be strongly chemorepellent (e.g. Tyr, Trp, Phe) amino acids.

Bacterial Chemotaxis

Some bacteria, such as *E. coli,* have several flagella per cell (4–10 typically). These can rotate in two ways :

- Counter-clockwise rotation aligns the flagella into a single rotating bundle, causing the bacterium to swim in a straight line.
- Clockwise rotation breaks the flagella bundle apart such that each flagellum points in a different direction, causing the bacterium to tumble in place.

The directions of rotation are given for an observer outside the cell looking down the flagella toward the cell.

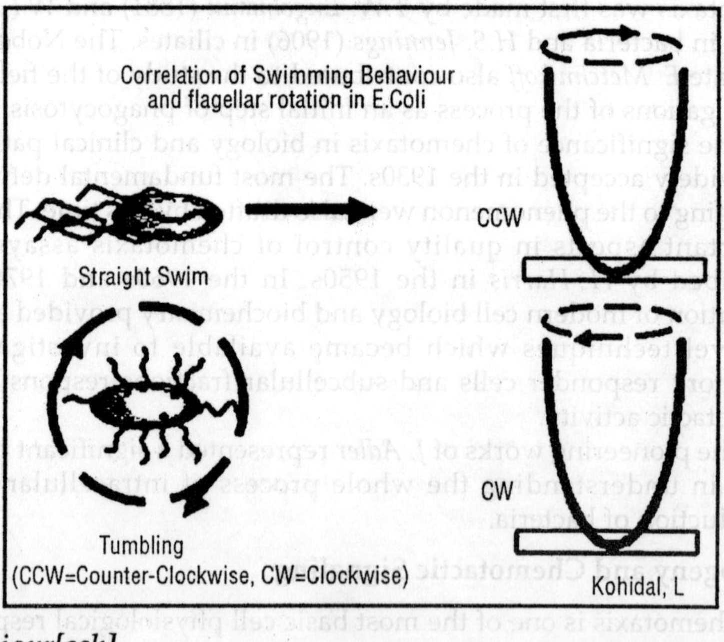

Behaviour[ssh]

The overall movement of a bacterium is the result of alternating tumble and swim phases. If one watches a bacterium swimming in a uniform environment, its movement will look like a random walk with relatively straight swims interrupted by random tumbles that reorient the bacterium.

Bacteria such as *E. coli* are unable to choose the direction in which they swim, and are unable to swim in a straight line for more than a few seconds due to rotational diffusion. In other words, bacteria "forget" the direction in which they are going. Given these limitations, it is remarkable that bacteria can direct their motion to find favorable locations with high concentrations of attractants (usually food) and avoid repellents (usually poisons).

In the presence of a chemical gradient bacteria will chemotax, or direct their overall motion based on the gradient. If the bacterium senses that it is moving in the correct direction (toward attractant/away from repellent), it will keep swimming in a straight line for a longer time before tumbling. If it is moving in the wrong direction, it will tumble sooner and try a new direction at random.

In other words, bacteria like *E. coli* use temporal sensing to decide whether life is getting better or worse. In this way, it finds the location with the highest concentration of attractant (usually the source) quite well. Even under very high concentrations, it can still distinguish very small differences in concentration. Fleeing from a repellent works with the same efficiency.

It seems remarkable that this purposeful random walk is a result of simply choosing between two methods of random movement; namely tumbling and straight swimming. In fact, chemotactic responses such as *forgetting* direction and *choosing* movements resemble the decision-making abilities of higher lifeforms with brains that process sensory data.

The helical nature of the individual flagellar filament is critical for this movement to occur. As such, the protein that makes up the flagellar filament, flagellin, is quite similar among all flagellated bacteria. Vertebrates seem to have taken advantage of this fact by possessing an immune receptor (TLR5) designed to recognize this conserved protein.

As in many instances in biology, there are bacteria that do not follow this rule. Many bacteria, such as *Vibrio*, are monoflagellated and have a single flagellum at one pole of the cell. Their method of chemotaxis is different. Others possess a single flagellum that is kept inside the cell wall. These bacteria move by spinning the whole cell, which is shaped like a corkscrew.

Signal Transduction

Chemical gradients are sensed through multiple transmembrane receptors, called methyl accepting chemotaxis proteins (MCPs), which vary in the molecules that they detect. These receptors may bind attractants or repellents directly or indirectly through interaction with proteins of periplasmatic space.

The signals from these receptors are transmitted across the plasma membrane into the cytosol, where *Che proteins* are activated. The Che proteins alter the tumbling frequency, and alter the receptors.

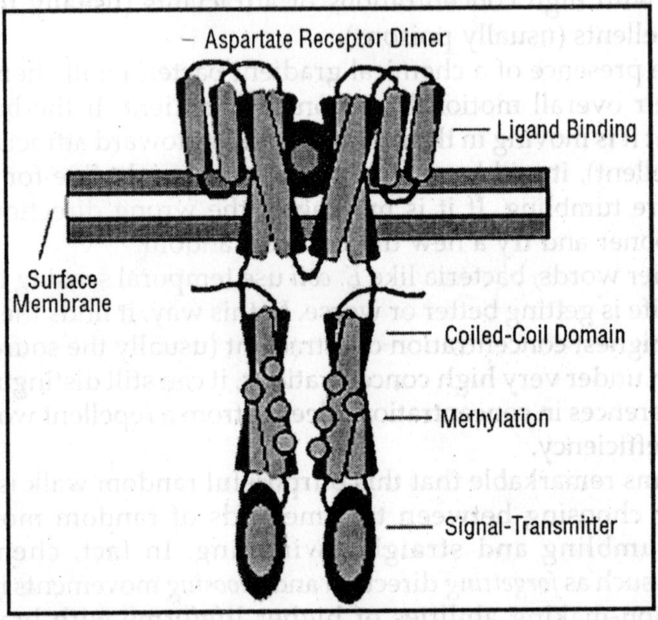

Flagellum Regulation

The proteins CheW and CheA bind to the receptor. The activation of the receptor by an external stimulus causes autophosphorylation in the histidine kinase, CheA, at a single highly conserved histidine residue. CheA in turn transfers phosphoryl groups to conserved aspartate residues in the response regulators CheB and CheY [note: CheA is a histidine kinase and it does not actively transfer the phosphoryl group.

The response regulator CheB takes the phosphoryl group from CheA]. This mechanism of signal transduction is called a two-component system and is a common form of signal transduction in bacteria. CheY induces tumbling by interacting with the flagellar switch protein FliM, inducing a change from counter-clockwise to clockwise rotation of the flagellum. Change in the rotation state of a single flagellum can disrupt the entire flagella bundle and cause a tumble.

Receptor Regulation

CheB, when activated by CheA, acts as a methylesterase, removing methyl groups from glutamate residues on the cytosolic side of the receptor. It works antagonistically with CheR, a methyltransferase, which adds methyl residues to the same glutamate residues.

The more methyl residues are attached to the receptor, the more sensitive the receptor. As the signal from the receptor induces demethylation of the receptor in a feedback loop, the system is continuously adjusted to environmental chemical levels, remaining sensitive for small changes even under extreme chemical concentrations.

This regulation allows the bacterium to 'remember' chemical concentrations from the recent past, a few seconds, and compare them to those it is currently experiencing, thus 'know' whether it is traveling up or down a gradient. However, the methylation system alone cannot account for the wide range of sensitivity that bacteria have to chemical gradients. Additional regulatory mechanisms such as receptor clustering and receptor-receptor interactions also modulate the signalling pathway.

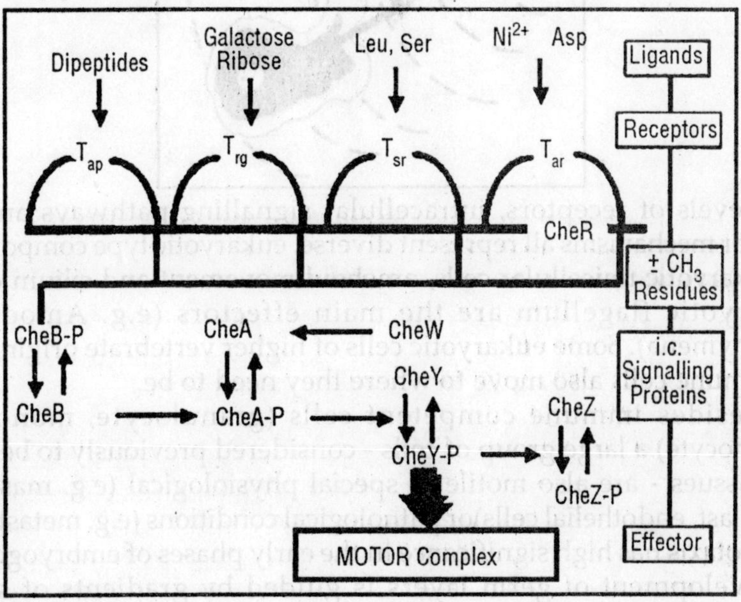

Eukaryotic Chemotaxis

The mechanism by which eukaryotic cells chemotax is quite different from that in bacteria; however, sensing of chemical gradients is still a crucial step in the process. Due to their size, prokaryotes cannot detect effective concentration gradients, therefore these cells scan and evaluate their environment by a constant swimming (consecutive steps of straight swims and tumbles).

In contrast to prokaryotes, the size of eukaryotic cells allows for the possibility of detecting gradients, which results in a dynamic and polarized distribution of receptors. Induction of these receptors by chemoattractants or chemorepellents results in migration towards or away from the chemotactic substance.

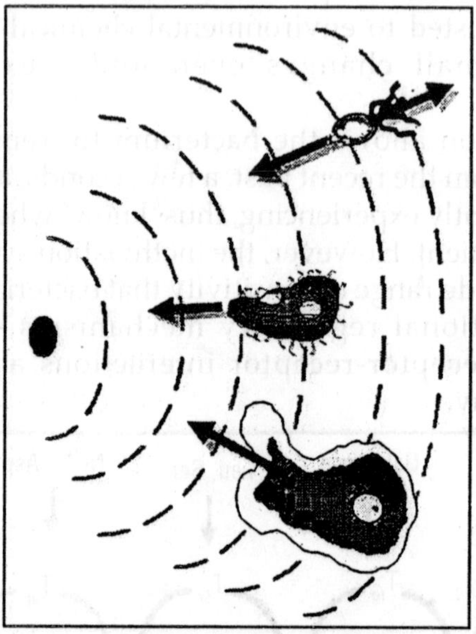

Levels of receptors, intracellular signalling pathways and the effector mechanisms all represent diverse, eukaryotic type components. In eukaryotic unicellular cells, ameboid movement and cilium or the eukaryotic flagellum are the main effectors (e.g. Amoeba or Tetrahymena). Some eukaryotic cells of higher vertebrate origin, such as immune cells also move to where they need to be.

Besides immune competent cells (granulocyte, monocyte, lymphocyte) a large group of cells - considered previously to be fixed into tissues - are also motile in special physiological (e.g. mast cell, fibroblast, endothelial cells)or pathological conditions (e.g. metastases). Chemotaxis has high significance in the early phases of embryogenesis as development of germ layers is guided by gradients of signal molecules.

Motility

Unlike motility in bacterial chemotaxis, the mechanism by which eukaryotic cells physically move is unclear. There appear to be mechanisms by which an external chemotactic gradient is sensed and turned into an intracellular PIP3 gradient, which results in a gradient and the activation of a signaling pathway, culminating in the polymerisation of actin filaments.

The growing distal end of actin filaments develops connections with the internal surface of the plasma membrane via different sets of peptides and results in the formation of pseudopods. Cilia of eukaryotic cells can also produce chemotaxis; in this case it is mainly a Ca^{2+} dependent induction of the microtubular system of the basal body and the beat of the 9+2 microtubules within cilia.

The orchestered beating of hundreds of cilia is synchronized by a submembranous system built between basal bodies. The details of the signaling pathways are still not totally clear.

CHEMOTAXIS RELATED MIGRATORY RESPONSES

Although chemotaxis is the most frequently studied form of migration there are several other forms of locomotion in the cellular level.

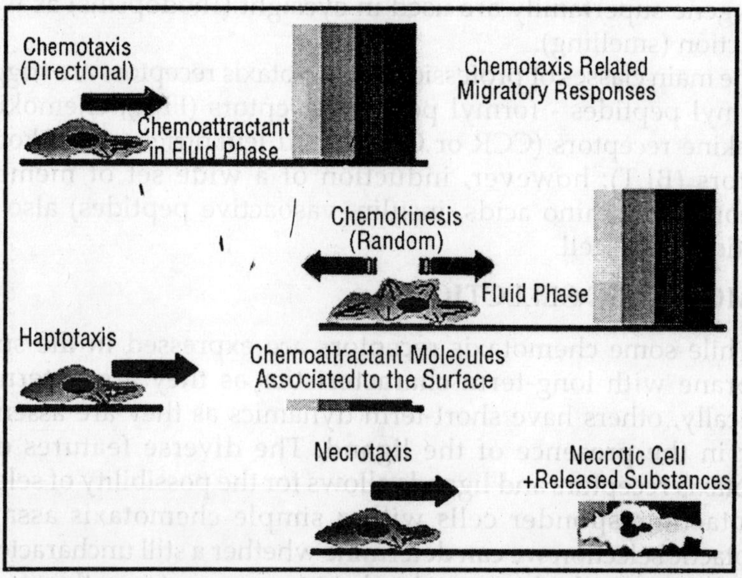

- *Chemokinesis* is also induced by molecules of the liquid phase of the surrounding environment; however, the response elicited is a not vectorial, random taxis. Neither amplitude nor frequency of motion has characteristic, directional components as this behaviour provides more scanning of the environment than migration between two distinct points.
- In *haptotaxis* the gradient of the chemoattractant is expressed

or bound on a surface, in contrast to the classical model of chemotaxis, in which the gradient develops in a soluble fluid. The most common biologically active haptotactic surface is the extracellular matrix (ECM); the presence of bound ligands is responsible for induction of transendothelial migration and angiogenesis.

- *Necrotaxis* embodies a special type of chemotaxis when the chemoattractant molecules are released from necrotic or apoptotic cells. Depending on the chemical character of released substances, necrotaxis can accumulate or repel cells, which underlines the pathophysiological significance of this phenomenon.

RECEPTORS

For the most part, eukaryotic cells sense the presence of chemotactic stimuli though the use of 7-transmembrane (or serpentine) heterotrimeric G-protein coupled receptors. This class of receptors is huge, representing a significant portion of the genome. Some members of this gene superfamily are used in eyesight (rhodopsins) as well as in olfaction (smelling).

The main classes of professional chemotaxis receptors are triggered by formyl peptides - formyl peptide receptors (FPR), chemokines - chemokine receptors (CCR or CXCR) and leukotrienes - leukotriene receptors (BLT); however, induction of a wide set of membrane receptors (e.g. amino acids, insulin, vasoactive peptides) also elicit migration of the cell.

CHEMOTACTIC SELECTION

While some chemotaxis receptors are expressed in the surface membrane with long-term characteristics as they are determined genetically, others have short-term dynamics as they are assembled *ad hoc* in the presence of the ligand. The diverse features of the chemotaxis receptors and ligands allows for the possibility of selecting chemotactic responder cells with a simple chemotaxis assay. By chemotactic selection we can determine whether a still uncharacterized molecule acts via the long- or the short-term receptor pathway.

The term *chemotactic selection* is also used to designate a technique which separates eukaryotic or prokaryotic cells according to their chemotactic responsiveness to selector ligands.

CHEMOTACTIC LIGANDS

The number of molecules capable of eliciting chemotactic

responses is relatively high, and we can distinguish primary and secondary chemotactic molecules. The main groups of the primary ligands are as follows:

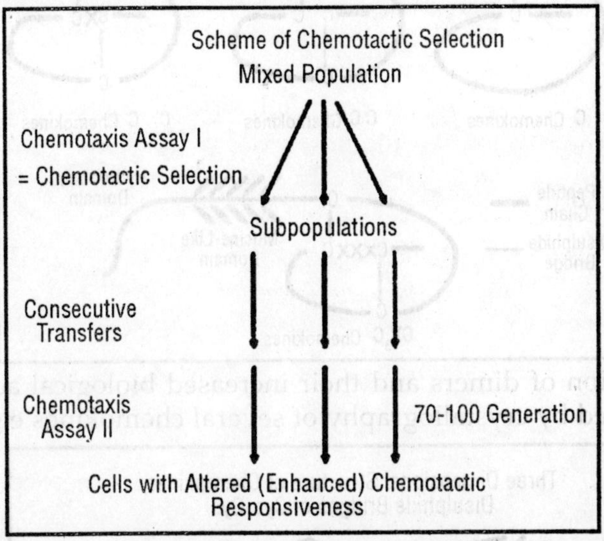

- *Formyl peptides* are di-, tri-, tetrapeptides of bacterial origin. They are released from bacteria in vivo or after decomposition of the cell. A typical member of this group is the N-formylmethionyl-leucyl-phenylalanine (fMLF or fMLP in references). The bacterial origin fMLF as a key component of inflammation has characteristic chemoattractant effects in neutrophil granulocytes and monocytes.
- *Complement 3a (C3a) and complement 5a (C5a)* are intermediate products of complement cascade. Their synthesis is joined to the three alternative pathways (classical, lectin dependent and alternative) of complement activation by a convertase enzyme. The main target cells of these derivaties are neutrophil granulocytes and monocytes as well.
- *Chemokines* belong to a special class of cytokines. Their groups represent not only structurally related molecules with a special arrangement of disulfide bridges, but their target cell specificity is also diverse: CC chemokines are acting on monocytes (e.g. RANTES), CXC chemokines are neutrophil granulocyte specific.

Investigations of the three-dimensional structures of chemokines proved that a characteristic composition of beta-sheets and an alpha helix provides expression of sequence s required for interaction with the chemokine receptors.

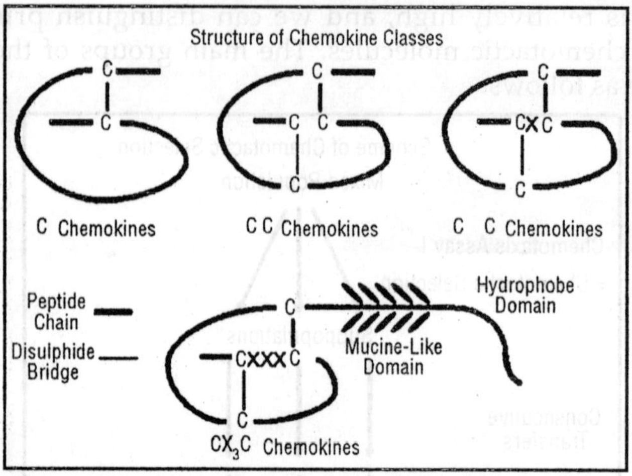

Formation of dimers and their increased biological acitvity was demonstrated by crystallography of several chemokines e.g. IL-8.

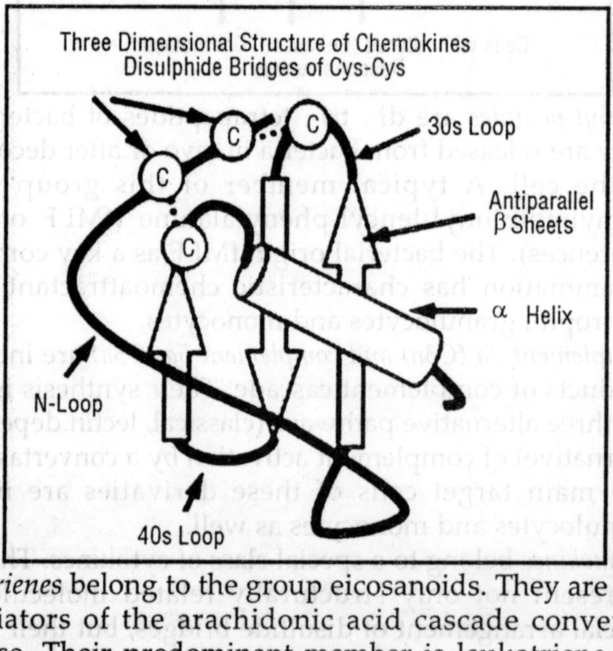

Leukotrienes belong to the group eicosanoids. They are significant lipid mediators of the arachidonic acid cascade converted by 5-lipoxigenase. Their predominant member is leukotriene B4 (LTB4) which elicits adhesion, chemotaxis and aggregation of leukocytes.

The characteristic chemoattractant effect of LTB4 is induced via G-protein linked seven-transmembrane spanning leukotriene receptors which are highly expressed in inflammation and allergy.

CHEMOTACTIC RANGE FITTING (CRF)

Chemotactic responses elicited by the ligand-receptor interactions are distinguished generally upon the optimal effective concentration(s) of the ligand.

Nevertheless, correlation of the amplitude elicited and ratio of the responder cells compared to the total number are also characteristic features of the chemotactic signaling.

Investigations of ligand families (e.g. amino acids or oligo peptides) proved that there is a fitting of ranges (amplitudes; number of responder cells) and chemotactic activities: chemoattractant moiety is accompanied by wide ranges, while chemorepellent character by narrow ranges.

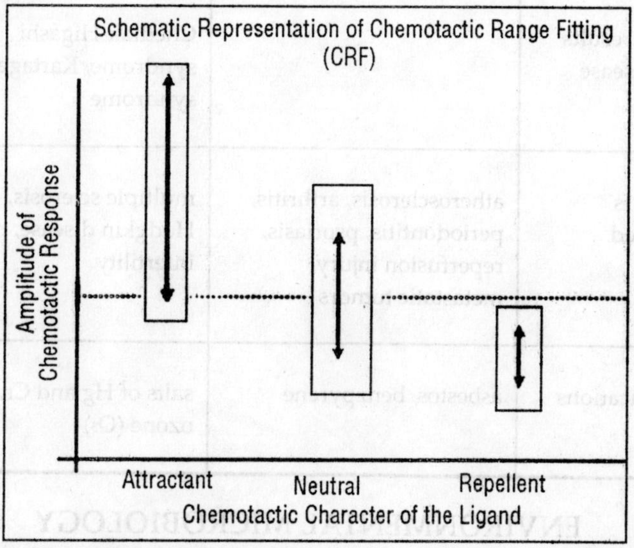

CLINICAL SIGNIFICANCE

A changed migratory potential of cells has relatively high importance in the development of several clinical symptoms and syndromes.

Altered chemotactic activity of extracellular (e.g. *Escherichia coli*) or intracellular (e.g. *Listeria monocytogenes*) pathogens itself represents a significant clinical target.

Modification of endogenous chemotactic ability of these microorganisms by pharmaceutical agents can decrease or inhibit the ratio of infections or spreading of infectious diseases.

Apart from infections, there are some other diseases where impaired chemotaxis is the primary etiological factor, as in Chediak-Higashi syndrome where giant intracellular vesicles inhibit normal migration of cells.

Chemotaxis (Chtx.) in diseases		
Type of disease	Chtx. increased	Chtx. decreased
Infections	inflammations	AIDS, Brucellosis
Chtx. results the disease	-	Chediak-Higashi syndrome, Kartagener syndrome
Chtx. is affected	atherosclerosis, arthritis, periodontitis, psoriasis, reperfusion injury, metastatic tumors	multiple sclerosis, Hodgkin disease, male infertility
Intoxications	asbestos, benzpyrene	salts of Hg and Cr, ozone (O_3)

ENVIRONMENTAL MICROBIOLOGY

Environmental microbiology is the study of the composition and physiology of microbial communities in the environment. The environment in this case means the soil, water, air and sediments covering the planet and can also include the animals and plants that inhabit these areas. Environmental microbiology also includes the study of microorganisms that exist in artificial environments such as bioreactors.

Microbial life is amazingly diverse and microorganisms literally cover the planet. It is estimated that we know fewer than 1% of the microbial species on Earth. Microorganisms can survive in some of the most extreme environments on the planet and some, for example the Archaea, can survive high temperatures, often above 100°C, as found in geysers, black smokers, and oil wells. Some are found in very cold habitats and others in highly saline, acidic, or alkaline water.

An average gram of soil contains approximately one billion

(1,000,000,000) microbes representing probably several thousand species. Microorganisms have special impact on the whole biosphere. They are the backbone of ecosystems of the zones where light cannot approach.

In such zones, chemosynthetic bacteria are present which provide energy and carbon to the other organisms there. Some microbes are decomposers which have ability to recycle the nutrients.

Microbes have a special role in biogeochemical cycles. Microbes, especially bacteria, are of great importance because their symbiotic relationship (either positive or negative) have special effects on the ecosystem. Microorganisms are used for *in-situ* microbial biodegradation or bioremediation of domestic, agricultural and industrial wastes and subsurface pollution in soils, sediments and marine environments.

The ability of each microorganism to degrade toxic waste depends on the nature of each contaminant. Since most sites typically have multiple pollutant types, the most effective approach to microbial biodegradation is to use a mixture of bacterial species and strains, each specific to the biodegradation of one or more types of contaminants. It is vital to monitor the composition of the indigenous and added bacteria in order to evaluate the activity level and to permit modifications of the nutrients and other conditions for optimizing the bioremediation process.

OIL BIODEGRADATION

Petroleum oil is toxic, and pollution of the environment by oil causes major ecological concern. Oil spills of coastal regions and the open sea are poorly containable and mitigation is difficult; much of the oil can, however, be eliminated by the hydrocarbon-degrading activities of microbial communities, in particular the hydrocarbonoclastic bacteria (HCB).

These organisms can help remedy the ecological damage caused by oil pollution of marine habitats. HCB also have potential biotechnological applications in the areas of bioplastics and biocatalysis.

DEGRADATION OF AROMATIC COMPOUNDS BY ACINETOBACTER

Acinetobacter strains isolated from the environment are capable of the biodegradation of a wide range of aromatic compounds. The predominant route for the final stages of assimilation to central metabolites is through catechol or protocatechuate (3,4-dihydroxybenzoate) and the beta-ketoadipate pathway and the diversity within the genus lies in the channelling of growth substrates, most of which are natural products of plant origin, into this pathway.

ANALYSIS OF WASTE BIOTREATMENT

Biotreatment, the processing of wastes using living organisms, is an environmentally friendly alternative to other options for treating waste material. Bioreactors] have been designed to overcome the various limiting factors of biotreatment processes in highly controlled systems.

This versatility in the design of bioreactors allows the treatment of a wide range of wastes under optimized conditions. It is vital to consider various microorganisms and a great number of analyses are often required.

ENVIRONMENTAL GENOMICS OF CYANOBACTERIA

The application of molecular biology and genomics to environmental microbiology has led to the discovery of a huge complexity in natural communities of microbes. Diversity surveying, community fingerprinting and functional interrogation of natural populations have become common, enabled by a range of molecular and bioinformatics techniques.

Recent studies on the ecology of Cyanobacteria have covered many habitats and have demonstrated that cyanobacterial communities tend to be habitat-specific and that much genetic diversity is concealed among morphologically simple types. Molecular, bioinformatics, physiological and geochemical techniques have combined in the study of natural communities of these bacteria.

CORYNEBACTERIA

Corynebacteria are a diverse group Gram-positive bacteria found in a range of different ecological niches such as soil, vegetables, sewage, skin, and cheese smear. Some, such as *Corynebacterium diphtheriae,* are important pathogens while others, such as *Corynebacterium glutamicum,* are of immense industrial importance. C. *glutamicum* is one of the biotechnologically most important bacterial species with an annual production of more than two million tons of amino acids, mainly L-glutamate and L-lysine.

LEGIONELLA

Legionella is common in many environments, with at least 50 species and 70 serogroups identified. *Legionella* is commonly found in aquatic habitats where its ability to survive and to multiply within different protozoa equips the bacterium to be transmissible and pathogenic to humans.

ARCHAEA

Originally, Archaea were once thought of as extremophiles existing only in hostile environments but have since been found in all habitats and may contribute up to 20% of total biomass. Archaea are particularly common in the oceans, and the archaea in plankton may be one of the most abundant groups of organisms on the planet. Archaea are subdivided into four phyla of which two, the Crenarchaeota and the Euryarchaeota, are most intensively studied.

LACTOBACILLUS

Lactobacillus species are found in the environment mainly associated with plant material. They are also found in the gastrointestinal tract of humans, where they are symbiotic and make up a portion of the gut flora.

WINOGRADSKY

In the late 1800s and early 1900s, Sergei Winogradsky, Russian microbiologist, pioneered the investigation of microbial auto trophy, and initiated the field of Environmental Microbiology. He was a strong supporter of examining freshly-isolated organisms rather than 'domesticated' laboratory strains.

One of the methods he developed for the study of microbial nutrient cycling in the environment is what is now known as the "Windogradsky column". These can be set up in an amazing variety of ways to study sulfur, nitrogen, carbon, phosphorus, or other nutrients, most often cycling between the upper aerobic zone and the lower anaerobic zone.

Chapter 4

Microbial Insecticides

Microbial insecticides are a new form of pesticide that work by infecting selected insect populations with bacteria, viruses, amoebas or fungi. Though this sounds potentially dangerous, many argue that it is actually quite safe, since the application of microbial insecticides is specific to the species one is trying to kill. Microbial insecticides usually have no effect on animal populations, unless diminishing a certain bug in the area interrupts the food chain. Each type usually works against only one type of insect.

Bacterial microbial insecticides may be used to control certain types of caterpillars that eat crops. They will kill caterpillars of both moths and butterflies, though, and should only be used where one will not diminish a butterfly population. Normally, this preparation is sprayed directly on crops.

One bacterial microbial insecticide works specifically on mosquito populations. It is considered extremely beneficial in eliminating populations that might spread the potentially deadly West Nile virus.

Several viral microbial insecticides work to first sicken and kill some insect species. They may affect moths, and sawflies, depending upon the virus used. Fungal microbial insecticides may be used on cockroaches, and create disease among a whole population. Amoebic insecticides may not kill an insect but may shorten the lifespan of an insect or cause it not to reach sexual maturity.

While microbial insecticides may be fantastic at killing a single type of insect, those with infestations of several different types of insects may require the use of several different sprays. Since microbial insecticides are so species specific, they are unlikely to harm any other bugs eating up or infesting crops, so they may not reduce all bug infestations at the same time.

Microbial insecticides also tend to be more vulnerable to outdoor elements. For example, long exposure to the sun, or heavy rains can kill certain bacteria. Therefore those attempting to control insect populations must be careful as to when it is appropriate to spray crops to achieve the maximum effect desired. Some scientists express concern over safety issues regarding the use of microbial insecticides.

We know for certain that living creatures even at the microscopic level change and evolve, as do the creatures they affect like insects. Could certain bugs become disease resistant, or could certain microbial insecticides mutate and affect other populations? These questions concern some environmentalists who view the widespread use of microbial insecticides as potentially dangerous in the future.

For now however, other environmentalists are celebrating the development of species specific agents that seem a better alternative to more frequently used poisonous pesticides. These environmentalists argue that microbial insecticides offer a way to get rid of harmful insects by keeping beneficial insects safe.

ADVANTAGES OF MICROBIAL INSECTICIDES

Individual products differ in important ways, but the following list of beneficial characteristics applies to microbial insecticides in general.

- The organisms used in microbial insecticides are essentially nontoxic and nonpathogenic to wildlife, humans, and other organisms not closely related to the target pest. The safety offered by microbial insecticides is their greatest strength.
- The toxic action of microbial insecticides is often specific to a single group or species of insects, and this specificity means that most microbial insecticides do not directly affect beneficial insects (including predators or parasites of pests) in treated areas.
- If necessary, most microbial insecticides can be used in conjunction with synthetic chemical insecticides because in most cases the microbial product is not deactivated or damaged by residues of conventional insecticides. (Follow label directions concerning any limitations.)
- Because their residues present no hazards to humans or other animals, microbial insecticides can be applied even when a crop is almost ready for harvest.
- In some cases, the pathogenic microorganisms can become established in a pest population or its habitat and provide control during subsequent pest generations or seasons.

DISADVANTAGES OF MICROBIAL INSECTICIDES

The limitations or disadvantages listed below do not prevent the successful use of microbial insecticides. Understanding how these limitations affect specific microorganisms will help users to choose effective products and take necessary steps to achieve successful results.

- Because a single microbial insecticide is toxic to only a specific species or group of insects, each application may control only a portion of the pests present in a field, garden, or lawn. If other types of pests are present in the treated area, they will survive and may continue to cause damage. Conventional insecticides are subject to similar limitations because they too are not equally effective against all pests. Nonetheless, the negative aspect of selectivity is often more noticeable for microbials.
- Heat, desiccation (drying out), or exposure to ultraviolet radiation reduces the effectiveness of several types of microbial insecticides. Consequently, proper timing and application procedures are especially important for some products.
- Special formulation and storage procedures are necessary for some microbial pesticides. Although these procedures may complicate the production and distribution of certain products, storage requirements do not seriously limit the handling of microbial insecticides that are widely available. (Store all pesticides, including microbial insecticides, according to label directions.)
- Because several microbial insecticides are pest-specific, the potential market for these products may be limited. Their development, registration, and production costs cannot be spread over a wide range of pest control sales. Consequently, some products are not widely available or are relatively expensive (several insect viruses, for example).

BACTERIA

Bacterial pathogens used for insect control are spore forming, rod-shaped bacteria in the genus *Bacillus*. They occur commonly in soils, and most insecticidal strains have been isolated from soil samples. Bacterial insecticides must be eaten to be effective; they are not contact poisons.

Insecticidal products comprised of a single *Bacillus* species may be active against an entire order of insects, or they may be effective against only one or a few species. For example, products containing *Bacillus thuringiensis* var. *kurstaki* kill the caterpillar stage of a wide array of butterflies and moths. In contrast, *Bacillus popillae* (milky spore disease) kills Japanese beetle larvae but is not effective against the closely related annual white grubs (masked chafers in the genus *Cyclocephala*) that commonly infest lawns.

The microbial insecticides most widely used in the United States since the 1960s are preparations of the bacterium *Bacillus thuringiensis* (abbreviated as *Bt*). *Bt* products are produced commercially in large industrial fermentation tanks. As the bacteria live and multiply in the right conditions, each cell produces (internally) a spore and a crystalline protein toxin called an endotoxin. Most commercial *Bt* products contain the protein toxin and spores, but some are cultured in a manner that yields only the toxin component.

How *Bt* Works

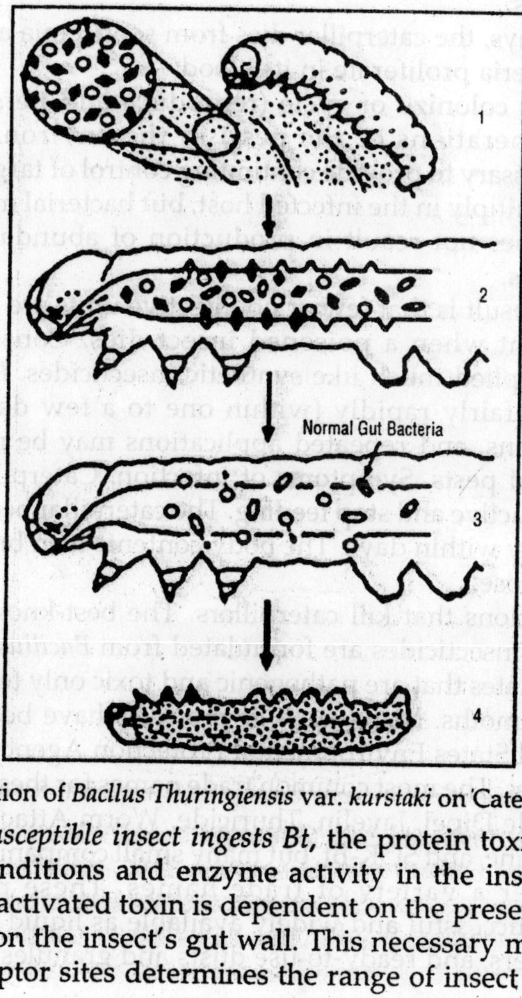

Fig. Action of *Baclus Thuringiensis* var. *kurstaki* on Caterpillars.

When *a susceptible insect ingests Bt,* the protein toxin is activated by alkaline conditions and enzyme activity in the insect's gut. The toxicity of the activated toxin is dependent on the presence of specific receptor sites on the insect's gut wall. This necessary match between toxin and receptor sites determines the range of insect species killed by each *Bt* subspecies and isolate. If the activated toxin attaches to receptor sites, it paralyzes and destroys the cells of the insect's gut wall, allowing the gut contents to enter the insect's body cavity and bloodstream. Poisoned insects may die quickly from the activity of the toxin or may die within 2 or 3 days from the effects of septicemia (blood poisoning). Although a few days may elapse before the insect dies, it stops feeding (and therefore stops damaging plants) soon after ingesting *Bt*.

- Caterpillar consumes foliage treated with Bt.
- Within minutes, the toxin binds to specific receptors in the gut wall, and the caterpillar stop feeding.
- Within hours, the gut wall breaks down, allowing spores and normal gut bacteria to enter the body cavity; the toxin dissolves.
- In 1-2 days, the caterpillar dies from septicemia as spores and gut bacteria proliferate in its blood.

Bt does not colonize or cycle (reproduce and persist to infect subsequent generations of the pest) in the environment in the magnitude necessary to provide continuing control of target pests. The bacteria may multiply in the infected host, but bacterial multiplication in the insect does not result in production of abundant spores or crystalline toxins.

The usual result is that few or no infective units are released into the environment when a poisoned insect dies. Consequently, *Bt* products are applied much like synthetic insecticides. *Bt* treatments are inactivated fairly rapidly (within one to a few days) in many outdoor situations, and repeated applications may be necessary for some plants and pests. Symptoms of infection. Caterpillars infected by *Bt* become inactive and stop feeding. The caterpillar becomes flaccid and dies, usually within days. The body contents turn brown to black as they decompose.

Bt formulations that kill caterpillars. The best-known and most widely used *Bt* insecticides are formulated from *Bacillus thuringiensis* var. *kurstaki* isolates that are pathogenic and toxic only to larvae of the butterflies and moths. Many such *Bt* products have been registered with the United States Environmental Protection Agency.

Trade names. The most common trade names for these commercial products include Dipel, Javelin, Thuricide, Worm Attack, Caterpillar Killer, Bactospeine, and SOK-Bt, but many small companies sell similar products under a variety of trade names. These products are commercially successful and widely available as liquid concentrates, wettable powders, and ready-to-use dusts and granules.

They are used to control many common leaf-feeding caterpillars, including caterpillar pests on vegetables (especially the "worms" that attack cabbage, broccoli, cauliflower, and Brussels sprouts), bagworms and tent caterpillars on trees and shrubs, larvae of the gypsy moth and other forest caterpillars, and European corn borer larvae in field corn. Several of these products are used to control Indianmeal moth larvae in stored grain, One product with a very specific target is Certan, formulated from *Bacillus thuringiensis* var. *aizawai*, and used exclusively for the control of wax moth larvae in honeybee hives.

Specificity. *Bt* products that kill caterpillars are NOT effective against other types of pests; they will not control aphids, beetles, flies, or additional pests other than caterpillars.

Even certain caterpillars are not effectively controlled by *Bt*, especially those that live in the soil or bore into plant tissues without consuming a significant amount of the *Bt* applied to plant surfaces. (Again, *Bt* is a stomach poison that must be ingested to be effective.)

The corn earworm in corn and the cutworms that clip off field crops or garden plants are examples of caterpillars seldom controlled by *Bt* treatments. Common caterpillar pests that are controlled effectively with *Bacillus thuringiensis* var. *kurstaki* (*Bt*) include:

- European corn borer in corn
- Indianmeal moth in stored grain
- Cabbage looper
- Imported cabbageworm
- Diamondback moth
- Tomato/tobacco hornworm
- Gypsy moth
- Spruce budworm
- Tent caterpillars
- Fall webworm
- Mimosa webworm
- Bagworms
- Spring and fall cankerworm

Common caterpillar pests that are NOT controlled by normal applications of *Bt* include:

- Corn earworm (on corn)
- Codling moth
- Peach tree borer
- Squash vine borer

Common caterpillar pests that are NOT controlled by normal applications of *Bt* include:

- Corn earworm (on corn)

- Squash vine borer
- Cutworms

USING Bt INSECTICIDES

Insecticides containing *Bt* can be very effective for insect control in a variety of situations. Reviewing a few key facts about these products can help users obtain the best results possible. Each *Bt* insecticide controls only certain types of insects; therefore, it is essential to identify the target pest and to confirm that the *Bt* insecticide controls only certain types of insects; therefore, it is essential to identify the target pest and to confirm that the *Bt* product label states that the insecticide is effective against that pest.

Separate stages of insects differ in their susceptibility to *Bt*; isolates that are effective against larval stages of butterflies or moths, will not kill adults. Because susceptible insects must consume *Bt* to be poisoned, treatments must be directed to the plant parts or other material that the target pest will eat. Where this is not possible (for example, where pests bore into plant tissue without feeding much on the surface foliage or fruits), *Bt* is usually not very effective.

Bt does not kill susceptible insects immediately. Poisoned insects normally remain on plants for a day or two after treatment, but they do not continue feeding and will die soon. Where *Bt* is applied to plant surfaces or other sites exposed to sunlight, it is deactivated rapidly by direct ultraviolet radiation. To maximize the effectiveness of *Bt* treatments, sprays should thoroughly cover all plant surfaces, including the undersides of leaves.

Treating in the late afternoon or evening can be helpful because the insecticide remains effective on foliage overnight before being inactivated by exposure to intense sunlight the following day. Treating on cloudy (but not rainy) days provides a similar result. Production processes that encapsulate *Bt* spores or toxins in a granular matrix (such as starch) or within killed cells of other bacteria also provide protection from ultraviolet radiation. Registration and sale of products containing encapsulated *Bt* are forthcoming.

Bt formulations that kill Mosquito, Black fly, and Fungus gnat Larvae

Production of a second group of *Bt* insecticides began in the early 1980s. These products utilize *Bacillus thuringiensis* var. *israelensis (Bti)*, a subspecies that kills the larvae of certain Diptera (the insect order containing the flies and mosquitoes). The main targets for *Bti* are the larval stages of mosquitoes, black flies, and fungus gnats; it does not control larval stages of "higher" flies such as the house fly, stable fly, or blow flies.

Mosquitoes that are most susceptible to *Bti* include species in the genera *Aedes* and *Psorophora. Anopheles* and *Culex* species are controlled only when higher than normal rates of *Bti* are applied. *Bti* products that are available commercially include Vectobac, Teknar, Bactimos, Skeetal, and Mosquito Attack.

Bti is most effective for mosquito or black fly control when it is used on a community-wide basis by mosquito abatement district personnel. In general, any attempts to reduce the prevalence of mosquito-breeding sites and treat remaining sites are most effective when undertaken on such a scale.

For most homeowners or farmers, eliminating sites that periodically serve as sources of standing water (such as tires and empty containers) and controlling weeds around ponds or drainage lagoons is more effective than applying *Bti* .

Bti products are formulated for spray or granular applications. *Bti* formulated on corn cob granules is effective against mosquito larvae developing in tires and other artificial containers. (The "Asian tiger mosquito," *Aedes albopictus,* develops in these containers.) The corn cob granules can be blown into tire piles to provide good penetration and uniform treatment; residual control is also greater when corn cob granules are used.

Bti is not very effective for the control of mosquito larvae in turbid water or waters containing high levels of organic pollutants.

Bti has also been used effectively for the control of fungus gnat larvae in greenhouses and in mushroom culture beds. For these uses, *Bti* is applied as a drench to potting soils or culture media. Not all *Bti* products are labeled for fungus gnat control.

Although not a *Bt* , another bacterium that is pathogenic to certain mosquitoes is *Bacillus sphaericus. Bacillus sphaericus* is especially active against larvae of mosquitoes in the genera *Culex, Psorophora,* and *Culiseta.* Effectiveness against larvae of *Aedes* species varies. *Aedes vexans* is very susceptible, but *Aedes aegypti* (the yellow fever mosquito) and *Aedes albopictus* are not.

Bacillus sphaericus kills *Anopheles* larvae in laboratory tests, but field results have not been promising. Because *Anopheles* mosquitoes are true surface feeders, the bacterium would have to remain on the water surface for an extended period to be effective. *Bacillus sphaericus* formulations tested up to now do not remain at the surface long enough to be effective.

In addition to infecting a different group of mosquito species than

Bti, Bacillus sphaericus is a potentially valuable insecticide because it remains effective in stagnant or turbid water. Although *Bacillus sphaericus,* as it occurs naturally, does cycle and maintain itself in the environment, insecticidal formulations currently under development do not cycle in water to infect subsequent generations of mosquito larvae.

Bt Formulations that kill Beetles

Another group of *Bt* isolates, including those from *Bacillus thuringiensis* var. *san diego* and *Bacillus thuringiensis* var. *tenebrionis,* are toxic to certain beetles. Within the order Coleoptera (the beetles), species exhibit great differences in susceptibility to these isolates, presumably because of differences in the insects' gut wall receptor sites where the bacterial toxins must attach. Consequently, the range of susceptible hosts for the beetle-targeted *Bt* formulations does not include all beetles, or even all of the species within a beetle family or subfamily.

In 1988, *Bacillus thurigiensis* var. *san diego,* sold under the trade name M-One, was first registered for use against larvae of the Colorado potato beetle. This product also kills elm leaf beetle adults and larvae, but it is not pathogenic or toxic to some other key beetle pests, such as the corn rootworms and other related species.

Registration of Trident, a very similar product containing *Bacillus thuringiensis* var. *tenebrionis,* is pending. Considerable research effort is now directed to identifying and developing additional *Bt* isolates that are active against more or different beetle species. Although entomologists and consumers alike will need to consider the specific target insect when judging the potential for these new products, *Bt* formulations effective against beetles seem to offer great promise.

Using Bt Insecticides

Insecticides containing *Bt* can be very effective for insect control in a variety of situations. Reviewing a few key facts about these products can help users obtain the best results possible. Each *Bt* insecticide controls only certain types of insects; therefore, it is essential to identify the target pest and to confirm that the *Bt* insecticide controls only certain types of insects; therefore, it is essential to identify the target pest and to confirm that the *Bt* product label states that the insecticide is effective against that pest.

Separate stages of insects differ in their susceptibility to *Bt;* isolates that are effective against larval stages of butterflies, moths, or mosquitos will not kill adults. Because susceptible insects must consume *Bt* to be poisoned, treatments must be directed to the plant parts or other material that the target pest will eat.

Where this is not possible (for example, where pests bore into plant tissue without feeding much on the surface foliage or fruits), *Bt* is usually not very effective. *Bt* does not kill susceptible insects immediately. Poisoned insects normally remain on plants for a day or two after treatment, but they do not continue feeding and will die soon.

Where *Bt* is applied to plant surfaces or other sites exposed to sunlight, it is deactivated rapidly by direct ultraviolet radiation. To maximize the effectiveness of *Bt* treatments, sprays should thoroughly cover all plant surfaces, including the undersides of leaves. Treating in the late afternoon or evening can be helpful because the insecticide remains effective on foliage overnight before being inactivated by exposure to intense sunlight the following day.

Treating on cloudy (but not rainy) days provides a similar result. Production processes that encapsulate *Bt* spores or toxins in a granular matrix (such as starch) or within killed cells of other bacteria also provide protection from ultraviolet radiation. Registration and sale of products containing encapsulated *Bt* are forthcoming.

Some *Bt* isolates (not those used in currently available insecticides) produce significant amounts of an additional toxin called thuringiensin, an exotoxin that is released outside the bacterial cell wall. Research is underway to develop commercial insecticides containing this toxin. Although thuringiensin might be lauded as "natural" because it is produced by living organisms, it is nonetheless toxic to a wide range of animal species and humans. As thuringiensin insecticides are registered and become available, users should recognize the difference between thuringiensin and other *Bt* products. Thuringiensin is much more toxic and should be handled much more cautiously.

Although the issue of thuringiensin's toxicity to mammals is a unique characteristic that does not detract from the overall safety of registered microbial insecticides, users are advised to handle all microbial insecticides cautiously. Bacterial spores, mold spores, and virus particles become "foreign proteins" if they are inhaled or rubbed into the skin.

As such, they can cause serious allergic reactions. The dusts or liquids used to dilute and carry these microorganisms also can act as allergens or irritants. These problems do not prevent the safe use of microbial insecticides, but they do mean that users should not breathe dusts or mists of microbial insecticides. Users should wear gloves, long sleeves, and long trousers during application and wash thoroughly after completing the application. These are common sense precautions that will help to prevent unexpected reactions and minimize any effects from unknown toxicity.

Recent advances in biotechnology have resulted not only in improved prospects for developing new *Bt* insecticides but also in an ability to place *Bt* toxins within crop plants in a variety of ways. For example, genes directing the production of *Bt* toxins can be incorporated into certain plant-dwelling bacteria. When these altered bacteria grow and multiply within an inoculated host plant, the *Bt* toxin is produced within the plant.

Efforts are underway to test this type of *Bt* "application" in corn to control the European corn borer. Although the development of this technology may seem ideal, the season-long, high-level control it would provide would also pose a great risk for the development of insect resistance to the *Bt* toxin. As genes for production of insecticidal compounds are added to crop plants, developers must also devise methods of preventing or managing insecticide resistance in target pests.

Other Bacterial Insecticides

Insecticides sold under the trade names Doom, Japidemic, Grub Attack, and the generic name "milky spore disease" contain the bacteria *Bacillus popillae* and *Bacillus lentimorbus.* These bacteria cannot be grown in fermentation tanks; instead, they are "cultivated" in laboratory-reared insect larvae. Products containing *Bacillus popillae* and *Bacillus lentimorbus* can be applied to turf and "watered in" to the soil below to control the larval (grub) stage of the Japanese beetle and, less effectively, some other beetle grubs.

When a susceptible grub consumes spores of these bacteria, they proliferate within it, and the grub's internal organs are liquefied and turned milky white (hence the name, milky spore disease). These symptoms develop slowly, often over a period of three to four weeks following initial infection.

Bacillus popillae and *Bacillus lentimorbus,* unlike *Bt,* do cycle in the environment if a substantial grub population is present at the time of application. When grubs killed by these bacteria break apart, a new batch of spores is released into the soil. These spores can survive (waiting to infect another grub) beneath undisturbed sod for a period of 15 to 20 years. Consequently, lawn applications of milky spore disease bacteria might not have to be repeated each year.

Unfortunately, *Bacillus popillae* and *Bacillus lentimorbus* offer limited usefulness in Illinois and other Midwestern states because the predominant lawn grubs in this region are annual white grubs, which are larvae of beetles called chafers (genus *Cyclocephala*). These larvae are not susceptible (or are only slightly susceptible) to milky spore disease.

Precautions

Users are advised to handle all microbial insecticides cautiously. Bacterial spores, as well as other microbes, become "foreign proteins" if they are inhaled or rubbed into the skin and can cause allergic reactions. The dusts or liquids used to dilute and carry these microorganisms also can act as allergens or irritants.

These problems do not prevent the safe use of microbial insecticides, but users should not breathe dusts or mists of microbial insecticides. Users should wear gloves, long sleeves, and long trousers during application and wash thoroughly afterwards. These are commonsense precautions that will help prevent unexpected reactions and minimize any effects from unknown toxicity.

VIRUSES

The larvae of many insect species are vulnerable to devastating epidemics of viral diseases. The viruses that cause these outbreaks are very specific, usually acting against only a single insect genus or even a single species.

Most of the viruses that are nuclear polyhedrosis viruses (NPV's), in which numerous virus particles are "packaged" together in a crystalline envelope within insect cell nuclei, or granulosis viruses (GV's), in which one or two virus particles are surrounded by a granular or capsule-like protein crystal found in the host cell nucleus. These groups of viruses infect caterpillars and the larval stages of sawflies.

Viruses, like bacteria, must be ingested to infect insect hosts. In sawfly larvae, virus infections are limited to the gut, and disease symptoms are not as obvious as they are in caterpillars. In caterpillars, virus particles pass through the insect's gut wall and infect other body tissues. As an infection progresses, the caterpillar's internal organs are liquefied, and its cuticle (body covering) discolors and eventually ruptures.

Caterpillars killed by viral infection appear limp and soggy. They often remain attached to foliage or twigs for several days, releasing virus particles that may be consumed by other larvae. The pathogen can be spread throughout an insect population in this way (especially when rain drops help to splash the virus particles to adjacent foliage) and by infected adult females depositing virus-contaminated eggs.

Dissemination of viral pathogens is deterred by exposure to direct sunlight, because direct ultraviolet radiation destroys virus particles. Although naturally occurring epidemics do control certain pests, these epidemics rarely occur before pest populations have reached out peak levels.

The development and use of virus-based insecticides have been limited. Unlike *Bt*, insect viruses must be produced in live host insects. Production is therefore both expensive and time-consuming.

Because viruses are genus- or species-specific, each viral insecticide has a limited market. These economic factors, coupled with the fact that some virus insecticides are considerably less effective than available synthetic chemical insecticides, have limited their development.

Nonetheless, although they are not well known or widely available, several insect viruses have been developed and registered for use as insecticides.

Most are specific to a single species or a small group of related forest pests, for example the gypsy moth, Douglas-fir tussock moth, spruce budworn, and pine sawfly.

They are not commercially available but are produced and used by the United States Forest Service.

Forest pests are especially good targets for viral pathogens because the permanence of the forest environment contributes to cycling of the pathogen (transmission from one generation to the next). The forest canopy also helps to protect viral particles from destruction by ultraviolet radiation.

Other insect viruses investigated for use as insecticides include those that infect the alfalfa looper, soybean looper, armyworms, cabbage looper, and imported cabbageworm. Although some of these viruses have been formulated and applied in field tests, none has been registered or sold commercially.

Both the codling moth GV (Decyde) and the *Heliothis* NPV (Elcar) were at one time registered by the US EPA and produced commercially, but these products are no longer registered or available.

If additional viral insecticides are registered or if currently registered products become more widely available, their effective use will depend upon applicators remembering the following key facts:

Most viruses are host-specific and effective only against immature stages of the target species; users must be sure to match the pathogen and the target pest correctly. Virus particles are killed by ultraviolet radiation; treating in the evening or on cloudy days should increase their effectiveness.

FUNGI

Fungi, like viruses, often act as important natural control agents that limit insect populations. Most of the species that cause insect diseases spread by means of asexual spores called conidia. Although conidia of different fungi vary greatly in ability to survive adverse environmental conditions, desiccation and ultraviolet radiation are important causes of mortality in many species.

Where viable conidia reach a susceptible host, free water or very high humidity is usually required for germination. Unlike bacterial spores or virus particles, fungal conidia can germinate on the insect cuticle and produce specialized structures that allow the fungus to penetrate the cuticle and enter the insect's body. Fungi do not have to be ingested to cause infections. In most instances, as fungal infections progress, infected insects are killed by fungal toxins, not by the chronic effects of parasitism.

Fungal pathogens differ in the range of life stages and species they are able to infect. Many important fungal pathogens attack eggs, immatures, and adults of a variety of insect species. Others are more specific to immature stages or to a narrow range of insect species.

Several factors have limited the development of fungal insecticides in the United States. Although fungal pathogens (at least some species) can be produced on artificial media, large-scale production of most pathogens has not yet been accomplished.

Precise production and storage conditions must be established and maintained to ensure that infective spores are produced and stored without loss of viability before they are applied. Once applied, pathogenic fungi often are effective only if environmental conditions are favorable; high humidity or rainfall usually is important. Where fungal pathogens are incorporated in soil to control below-ground pests, the adverse effects of ultraviolet radiation and desiccation are minimized, but other microorganisms that act as competitors or antagonists often alter pathogen effectiveness.

Although no fungal pathogens are currently registered or commercially available in the United States, one or more fungi are used in Great Britain, China, the Soviet Union, and additional countries in eastern Europe and South America. Fungi used as insecticides include the following:

- *Beauveria bassiana:* This common soil fungus has a broad host range that includes many beetles and fire ants. It infects both larvae and adults of many species. One or more companies currently are developing preparations of this fungus for EPA registration. Understanding the interactions between *Beauveria bassiana* and other soil microorganisms may be the key to successful use of this fungus.

- *Nomuraea rileyi:* In soybeans (especially in the southeastern United States), naturally occurring epidemic infections of *Nomuraea rileyi* cause dramatic reductions in populations of foliage-feeding caterpillars. Research directed at predicting disease outbreaks caused by this fungus may help in determining the need for application of insecticides.
- *Vericillium lecanii* : This fungus (once sold under the trade name Vertelec) has been used in greenhouses in Great Britain to control aphids and whiteflies.
- *Lagenidium giganteum:* This aquatic fungus is highly infectious to larvae of several mosquito genera. It cycles effectively in the aquatic environment (spores produced in infected larvae persist and infect larvae of subsequent generations), even when mosquito density is low. Its effectiveness is limited by high temperatures.
- *Hirsutella thompsonii:* Although preparations of this pathogen were once registered by the US EPA and marketed under the trade name Mycar, it is no longer available commercially. *Hirsutella thompsonii* is a pathogen of the citrus rust mite.

PROTOZOA

Protozoan pathogens naturally infect a wide range of insect hosts. Although these pathogens can kill their insect hosts, many are more important for their chronic, debilitating effects. One important and common consequence of protozoan infection is a reduction in the number of offspring produced by infected insects. Although protozoan pathogens play a significant role in the natural limitation of insect populations, few appear to be suited for development as insecticides.

Species in the genera *Nosema* and *Vairimorpha* seem to offer the greatest potential for use as insecticides. Pathogens in these genera attack Lepidopteran larvae and insects in the order Orthoptera (the grasshoppers and related insects).

The one protozoan currently available in a registered insecticidal formulation is the microsporidian *Nosema locustae,* a pathogen of grasshoppers. It is sold as NOLO Bait and Grasshopper Attack. It is most effective when ingested by immature grasshoppers (the early nymphal stages). Infections progress slowly; where the pathogen kills the grasshopper, death occurs three to six weeks after initial infection. Not all infected hoppers are killed.

Nosema locustae has been used to reduce grasshopper populations in rangeland areas, and adequate control has been achieved when treatments were applied to large areas while hoppers were still young. Although not all grasshoppers in the treated area are killed by *Nosema locustae* , infected hoppers consume less forage and infected females produce fewer viable eggs than do uninfected females.

Nosema locustae persists on egg pods to provide varying degrees of infection the following season. The effectiveness and utilization of *Nosema locustae* for rangeland grasshopper control are likely to increase as research continues.

Unfortunately, small, one-pound packages of *Nosema locustae* preparations developed for sale to gardeners and homeowners offer much less utility (or none). The mobility of grasshoppers, coupled with the fact that infected hoppers are not killed until a few weeks after they ingest the pathogen, means that application of baits containing *Nosema locustae* to individual lawns or gardens is unlikely to reduce grasshopper densities or damage substantially.

(Knowing that we infected some grasshoppers that may or may not die later in someone else's garden may be somewhat satisfying, but it doesn't help very much if our goal was to prevent damage to our vegetables.)

NEMATODES

To be accurate, nematodes are not microbial agents. Instead, they are multicellular roundworms. Nematodes used in insecticidal products are, however, nearly microscopic in size, and they are used much like the truly microbial products discussed previously. Nematodes used for insect control infect only insects or related arthropods; they are called entomogenous nematodes.

The entomogenous nematodes *Steinernema feltiae* (sometimes identified as *Neoaplectana carpocapsae*), *S. scapteriscae*, *S. riobravis*, *S. carpocapsae* and *Heterorhabditis heliothidis* are the species most commonly used in insecticidal preparations. Within each of these species, different strains exhibit differences in their abilities to infect and kill specific insects. In general, however, these nematodes infect a wide range of insects. On a worldwide basis, laboratory or field applications have been effective against over 400 pest species, including numerous beetles, fly larvae, and caterpillars.

The infectious stage of these nematodes is the third juvenile stage often referred to as the J3 stage or the "dauer" larvae. Nematodes in this stage survive without feeding in moist soil and similar habitats, sometimes for extended periods. *Steinernema* species infect host insects by entering through body openings—the mouth, anus, and spiracles (breathing pores). *Heterorhabditis* juveniles also enter host insects through body openings, and in some instances are also able to penetrate an insect's cuticle. If the environment is warm and moist, these nematodes complete their life cycle within the infected insect. Infective juveniles molt to form adults, and these adults produce a new generation withing the same host. As the offspring mature to the J3 stage, they are able to leave the dead insect and seek a new host.

Symbiotic bacteria associated with entomogenous nematodes actually cause the death of infected insects. Once inside an insect host, nematodes empty their guy bacteria into the insect's blood stream. The nematodes then feed on these bacteria as they multiply. The insect dies from bacterial septicemia. These entomogenous nematodes and their associated bacteria (*Xenorhabdus luminescens*) have been tested extensively for toxicity to nontarget organisms, and they are considered to be nontoxic and nonpathogenic to plants and mammals.

Large-scale production of nematodes has been initiated, and a few insecticides containing entomogenous nematodes are now marketed. These products contain infective juveniles and are formulated for application as sprays or drenches.

Although *Steinernema* and *Heterorhabditis* species have been shown to be effective against a great number of insect species in laboratory experiments, their host range in field applications is often limited by temperature and moisture conditions in the target insect's habitat. Although controlled desiccation has been used successfully for the formulation of some insecticidal products containing nematodes, desiccation in the field usually kills entomogenous nematodes.

Consequently, nematodes are most effective when applied to control insects in moist soil, plant tissues, or other protected habitats. Their viability and persistence in soil varies greatly depending upon soil moisture and temperature and other organisms present. The following examples illustrate situations in which nematodes are likely to provide effective insect control.

For the control of annual white grubs in grass sod, entomogenous nematodes should be effective when applied at the right time to adequately watered turf. If applications are made when grubs are present (early to late August in Illinois, with correct timing dependent upon latitude and annual variations in weather), nematodes (and synthetic insecticides) need to remain active only for a period of several days to reduce grub damage.

Because most homeowners and turf managers are able to irrigate, they can effectively "water in" nematode applications and keep soils moist enough to favour nematode survival and grub infection.

Although nematodes may survive for extended periods (up to several months) in cool, moist soils, homeowners should probably expect that nematodes will persist for only two to four weeks following application for grub control.

Nematodes also have been used effectively in trials against root weevil larvae and similar soil pests attacking nursery plants, ornamental plantings, garden crops, and potted plants. The ability to maintain soil moisture is an important factor in the success of these treatments. Nematodes are being used to control the tawny and Aouthern mole crickets present in turf and pastures.

Nematac can also persist for many years to give long term supression. *Steinernema scapterisci* is a naturally occuring nematode that seeks out mole crickets. Nematodes have been used to effectively control flea larvae living in lawns and outdoor areas accessible to pets. The nematodes effectively kill flea larvae and have no toxicity to pets or humans.

Nematodes are not appropriate for termite control. Entomogenous nematodes are infectious to termites, but they are not likely to provide the long-term persistence needed in a termiticide. In soils around and under buildings, low moisture levels and the probable long-term absence of host insects would, in most cases, prevent nematode survival (persistence).

Consequently, the need for long-term control would require repeated applications of nematodes at intervals not yet determined. Because repeated applications are impractical and undesirable, nematodes are not recommended for termite control. Until further research provides data on specific pests, we recommend against the use of nematodes to control most above-ground insects.

When nematodes can be placed in moist, protected environments where insects are confined (such as tunnels bored into tree trunks or other plant tissues), treatments can be effective. When nematodes are applied to exposed foliage or other locations where desiccation occurs rapidly, treatments are not likely to be effective.

Although commercial availability is likely to increase, nematodes are currently sold primarily by mail order from gardening supply houses (or wholesale to large-volume purchasers). Trade names include Vector, Scanmask andBioSafe; some products simply use the scientific name of the nematode. Nematodes are not considered by the US EPA to be pathogens, and nematode producers are not required to complete the same EPA registration process required of producers of other microbial pesticides.

This exemption means reduced start-up costs for the manufacture of products containing entomogenous nematodes, but it also means less standardization and less outside regulation of quality. As major producers of entomogenous nematodes gain a reputation for marketing high-quality products, consumers can expect these companies to practice effective quality control.

Nonetheless, in the absence of extensive regulation, some suppliers are likely to market inferior products accompanied by exaggerated claims. Because of differences among nematode products and lack of regulations, buyers are urged to be careful consumers. These products offer a great deal of potential, but their efficacy is not as unlimited as some promotions suggest.

INSECT PARASITIC NEMATODES

Insect parasitic nematodes are small, round worms that complete part of their life cycle in insects. Several species can kill insects in this process, and some are marketed as a biological control agent.The use of insect parasitic nematodes to control soil insects, including many turfgrass and Garden pests, has received increased attention in recent years. insect parasitic nematodes are also know as predator nematodes, beneficial nematodes or entomogenous nematodes. They are sold under such trade names as BioSafe, BioVector, Scanmask, Exhibit, Oti-Nem, and Guardian.

Insect parasitic nematodes have been studied for potential biological control of turf pests for more than 50 years. However, their increasing availability and rising public demand for alternatives to insecticides has heightened interest in parasitic nematodes.

Nematodes are a type of roundworm in the phylum Nematoda. This is a different order of animals from the segmented worms, such as earthworms and flatworms.

Nematodes are abundant, particularly in soil. Many feed on bacteria, some attack plants, and others can be parasites of animals. However, each species is usually specific in its habits. insect parasitic nematodes, for example, develop only in insects or other arthropods. They do not attack mammals, birds or plants.

Two genera of insect parasitic nematodes have been most actively considered for control of insect pests. Steinernema (previously known as Neoaplectana) has been thoroughly researched, including control studies for Japanese beetle conducted in the 1930s. Recent attention has shifted to include the genus Heterorhabditis. Both groups are in the order Rhabdita, the bacteria-feeding nematodes. It includes a large number of common but frequently overlooked soil-infesting nematodes that assist in organic matter breakdown.

Life Cycle

Insect parasitic nematodes are mobile and move short distances in search of host insects. Heterorhabditis is more mobile than Steinernema. Nematodes use carbon dioxide and perhaps other chemicals produced in waste products of insects as cues to find their hosts. Upon entering the insect the nematodes kill their host by releasing bacteria into it, strains of Xenorhabdus species that are associated with the nematode.

The bacteria develop within the body cavity of the susceptible host. The host dies from blood poisoning within a few days.The dead insect generally maintains its original shape and does not decay in a normal manner because its body is filled with these specialized bacteria. Associated colour changes may occur. For example caterpillars parasitized by Heterorhabditis may have a reddish-brown colour. The nematodes develop by feeding upon the bacteria and degraded host tissues.

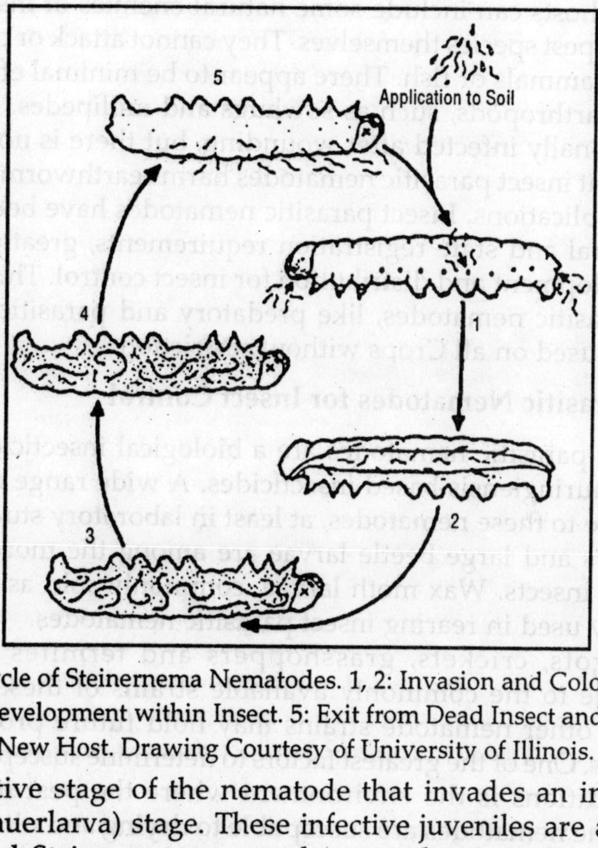

Fig. Life Cycle of Steinernema Nematodes. 1, 2: Invasion and Colonization of Insect. 3,4: Development within Insect. 5: Exit from Dead Insect and Invasion of New Host. Drawing Courtesy of University of Illinois.

The active stage of the nematode that invades an insect is the juvenile (dauerlarva) stage. These infective juveniles are applied for insect control. Steinernema nematodes enter the insect through natural openings, such as the mouth, spiracles and anus, then penetrate into the body cavity. Heterorhabditis nematodes use natural openings but also can enter by piercing the body wall. Once inside the infected insect, nematodes develop rapidly, in five or more days, depending on host quality and temperature. Normally, one to two generations occur in a host insect. Thousands of nematodes can be produced following a single infection. Ultimately, the body wall of the dead host insect ruptures and releases the nematodes. insect parasitic nematodes typically kill their host insect within two to three days after invading the body cavity.

Safety and Environmental Concerns

Insect parasitic nematodes do not appear to have any significant harmful effects on other beneficial organisms or people. Arthropods appear to be the only hosts for these organisms, so plants and most wildlife are not likely to be directly affected by nematode applications. However, hosts can include some natural enemies of insect pests, as well as the pest species themselves. They cannot attack or cause disease in birds, mammals or fish. There appear to be minimal effects against noninsect arthropods, such as sowbugs and millipedes. Earthworms are occasionally infected after wounding, but there is no evidence to suggest that insect parasitic nematodes harms earthworm populations in field applications. Insect parasitic nematodes have been exempted from federal and state registration requirements, greatly facilitating their development and distribution for insect control. This means that insect parasitic nematodes, like predatory and parasitic insects, can legally be used on all Crops without restriction.

Insect Parasitic Nematodes for Insect Control

Insect parasitic nematodes are a biological insecticide similar to Bacillus thuringiensis-based insecticides. A wide range of insects are susceptible to these nematodes, at least in laboratory studies. Various caterpillars and large beetle larvae are among the more susceptible groups of insects. Wax moth larvae, commonly sold as fish bait, are commonly used in rearing insect parasitic nematodes.

Maggots, crickets, grasshoppers and termites appear less susceptible to the commonly available strains of these nematodes. However other nematode strains may hold future promise against these pests. One of the greatest factors to determine susceptibility under field conditions is the environment where the pest insect occurs. Because the nematodes are susceptible to drying and ultraviolet light, they are most effective against insects that occur in moist, dark locations. This includes various soil insects or borers within woody plants. Insects on foliage may be susceptible in laboratory settings but control is poor in field applications.

Insect parasitic nematodes are sold in the infective juvenile dauerlarva stage, which are barely visible to the unaided eye. They typically are used at rates of 250 million to 2 billion per acre (approximately 6,000 to 46,000 per square foot). To apply, dilute the nematodes in water and drench the soil or inject them into plants. Regular spray equipment can be used — nematodes are quite tolerant of pressures found in many sprayers.

Steinernema

Steinernema is the most widely researched species for insect control. It is the most readily available for yard and Garden use because it is easier to rear and handle. In field applications, *Steinernema carpocapsae* tend to be most effective against caterpillar larvae.

In laboratory and field trials, it has controlled sod webworms, cutworms and certain borers (raspberry crown borer, carpenter worm). It also has been effective against billbug larvae in Colorado State University trials. Other research indicates that adult billbugs may be controlled as well. Another speices, *Steinernema feltiae* is effective against certain fly larvae. It is sometimes used to control fungus gnats and other soil dwelling insects of greenhouse-grown plants.

Steinernema are less effective against white grubs, root maggots, rootworms and black vine weevil. Unfortunately, some commercial products make claims of effective control of some pest species based on research conducted solely in artificial environments, these often do not reflect performance in the field.

Heterorhabditis

Heterorhabditis is less commonly available because it is more difficult to rear and more susceptible to environmental extremes. However, field trials consistently show that this genus outperforms Steinernema for control of white grubs. Several species are sold including *Heterorhabditis bacteriophora, H. megadis,* and *H. indica.*

Heterorhabditis also effectively controls many nursery pests that feed in the root zone, such as black vine weevil and citrus-infesting root weevils.

EFFECTS OF ENVIRONMENT ON PERFORMANCE OF NEMATODES

Effective use of nematodes for insect control depends on

environmental conditions during and after application. Adequate moisture is the most important factor. Insect parasitic nematodes require free water to move and are susceptible to death by drying. High humidity and free water are the primary conditions for effective use of insect parasitic nematodes.

Insect parasitic nematodes also are susceptible to extreme temperature conditions. For example, short exposure to high temperatures (above 90 to 95 degrees F) can be lethal to commonly available strains of *Steinernama carpocapsae*. Low temperatures (below about 55 degrees) inactivate most *Heterorhabditis* species. Strain and species differ in temperature sensitivity.

In soil applications, nematode movement to the target pest depends on conditions of the soil and thatch. Adequate moisture is essential and pretreatment irrigation is recommended. Irrigation may also cause soil insects, such as white grubs, to move higher in the soil where they may be more easily found by the nematodes.

Nematode movement is more extensive in sandy soils with large pore size than in clay soils. Nematode penetration may be expected to be poor in "tight" soils common to some regions. Extensive thatch accumulation may also prevent nematode penetration into the turf root zone where white grubs feed. A post-treatment irrigation may help push nematodes through the thatch layer.

Insect parasitic nematodes appear compatible with many chemicals used in lawn and Garden care. Nitrogen fertilizers are nontoxic to *Steinernema carpocapsae*, as are insecticidal soaps and wetting agents. Insecticides show variable effects on nematode survival, with most having little effect. For example, diazinon appears to be nontoxic to insect parasitic nematodes.

The carbamate insecticides carbaryl (Sevin) and bendiocarb (Turcam) appear to be moderately toxic, particularly to *Heterorhabditis*. Among tested turfgrass insecticides, chlorpyrifos (Dursban) is one of the most toxic to insect parasitic nematodes. Mercurial fungicides are even more damaging to insect parasitic nematodes, and effects are persistent.

STORAGE AND HANDLING

Most insect parasitic nematodes packaged for sale have a shelf life of several weeks or even months if refrigerated. *Steinernema carpocapsae* is particularly tolerant of storage conditions, a major reason for its ready availability. Systems to dry *Steinernema* nematodes for extended storage have been developed, a process that can increase storage life and reduce the need for refrigeration.

Regardless, insect parasitic nematodes can be killed or inactivated by extreme temperatures. Never allow them to freeze or be exposed to high temperatures (above about 90 degrees). Information is still lacking regarding the persistence of insect parasitic nematodes in turfgrass under Colorado conditions. Native nematodes do occur in the state.

One of the most spectacular is *Mermis nigrescens,* often over 4 inches long, which is a common parasite of grasshoppers. Commercially available strains have originated from the southern United States or other mild climates. Those that reproduce can produce multiple cycles of infection during the growing season. However, it is likely that extremes of cold, drying or high summer temperatures would kill off nematodes under Colorado conditions.

Furthermore, various soil organisms, including predatory mites, feed on nematodes and may reduce populations to ineffective levels. In addition, there is some dispersal of insect parasitic nematodes either through their own movements or by insects and mites. At present, there are no reliable guidelines on nematode persistence.

BENEFICIAL NEMATODES

Beneficial nematodes are microscopic worm-like organisms found naturally in soil. There are many species—good, bad and in-between. Good nematodes kill pests in soil. They work best when soil stays moist, killing in about 48 hours by releasing unique bacteria inside pest larva, hence they are called infective parasitic nematodes.

The following species are available commercially in various size packaging and formulations and will attack a broad host range, but use the following guide for which nematode species is best for different systems and pests. Heterorhabditis bacteriophora (also Nemasys G) can be used in lawn and field conditions in soils at 55º F to 86º F against:

- Cucumber, flea, scarab and Japanese beetles, strawberry root and black vine weevil, chafers, white grub, corn root worm, billbug, Colorado potato beetle, Oriental and Japanese beetles.
- Highly motile, and "cruising" in search of host, thus valuable when the pest is more dispersed.
- More potent, because they have a "tooth" to rupture and enter the insect's skin.
- Twice as good against white grubs compared to S. carpocapse.

Heterorhabditis marilatus more cold hardy, use on cranberries, black vine weevil, may be more virulent than H. bacteriorphora, inconsistently available, especially in summer. Heterorhabditis megidis (Nemasys H) is applied April-June and Aug-Nov for black vine weevil when soils are 55° F to 86° F. Heterorhabditis indica more heat tolerant and may be more virulent than H. bacteriorphora, inconsistently available, especially in winter. Steinernema carpocapse (also Millenium) are used in lawn, garden and orchard soils at 55° F to 90° F at 1 billion per acre (less for caterpillars and more for cranberry girdler, mint pests and weevils):

- Fleas, codling moth, navel orangeworm, peach tree borer, mint root borer, cranberry girdler, cutworm, armyworm, leafminer, sod web worm, mole cricket, banana moth, other caterpillar pests, termites, bluegrass weevils.
- S. carpocapse wait to ambush a passing host, latch on and crawl through body openings.
- S. carpocapse is the preferred species when hosts are mobile.
- Steinernema feltiae (also Nemasys) are used in media at 50° F to 86° F or as a foliar against:
- Fungus gnat, shore fly, mushroom gnats, fruit fly, Western flower thrips, leafminer.
- Biocontrol inhibition of some plant parasitic nematodes, particularly root-knot nematodes.
- House flies need 5-10 million nematodes per 25 sq. ft. of manure pile.

For high levels of fungus gnats, Steinernema move through rockwool, peat, sawdust or soil growing media without difficulty. They will not reproduce or stay in the growing media and must be reapplied if reinfestation occurs.

Follow these steps:

- Apply nematodes when yellow sticky traps average over 50-75 fungus gnat adults/trap, weekly, or while fungus gnat populations appear to be increasing.
- If root diseases are a problem in the crop, apply nematodes sooner to reduce the risk of disease transmission by fungus gnat larvae.
- Three applications, 7-10 days apart, are usually required. Follow product label for rates.
- Calculate based on actual growing area (i.e., containers or beds) not greenhouse area.

Steinernema kraussei (also Nemasys L) are applied June-Sept in turf at 41º F to 86º F against Japanese beetle, European chafer, Masked chafer, Oriental beetle, and black vine weevil. 250 million will treat 5500 square feet. Steinernema riobrave (BioVector) are used at soil temperatures of 50º F to 86º F or as a foliar:

- In citrus against citrus weevil, blue green weevil, borers at 20-

40,000 per square foot, 0.5 to 1 billion per acre for fuller rose beetle.

- In turf for mole crickets (or use Steinernema scapterisci, Nematac S).
- Black vine weevil and strawberry weevil at 3 billion per acre.
- Competes with root knot and sting nematodes.

Quantity Guidelines

In general beneficial nematodes supplied by Rincon-Vitova are used at around 1 million per 50 square feet and 1 billion per acre, but product labels vary. Release every three to six weeks until infestation subsides. The warmer and more regular the watering system, the lower the required rate. With lower rates, control declines. For example, using H. bacteriophora against Japanese beetle grubs, after three weeks, a billion per acre achieved 85-90% control while half that much resulted in half the control.

Spot Treatments Effective for Many Pests

For control of fleas, applications should be made to areas, which the pet frequents such as feeding and watering sites, beds, or shaded areas. When treating tree pests, treat the drip line area, about one-fifth of the total acreage. For tree bark pests, nematode solutions are injected into borer holes.

When and How to Apply

Soil temperature should be greater than 45º F to apply all species and warmer for some. At that temperature they will live but not grow very well. When soil reaches 60º F all species will grow and parasitize insects. Late evening or night is ideal and never when the ground is hot and dry. Water dry soils before applying. The best conditions are cloudy, cool weather in spring or late summer while the soil temperatures are below 85º but above 60º F. When using a sprayer, use a course nozzle, remove screens and pressures can be set up to 300 psi. The mixture can also be distributed through an irrigation system.

How it Comes

Our packaging varies from a moist sponge sealed in a plastic bag, water dispersible granules in trays or in vermiculite or clay particles. Pouches contain a water dispersible granule that we just mix with water. Clay particles (DE) are mixed with water, then the clay falls to the bottom of the container and the nematode suspension poured off. Vermiculite is sprinkled on the soil surface.

Directions for Mixing

Mix nematodes with water so we can sprinkle or spray them on the ground. Mix only what we will apply today. Use suspended nematodes right away. They will eventually drown in the water. The suspended nematodes survive about three to four hours depending on temperature, oxygen in the water and some other factors.

To check viability, examine a drop of the suspension with a hand lens. Live nematodes will be curved and ideally moving in the water drop. If dead, they are straight. To mix only half the quantity from a sponge, cut the smear on the sponge in half, use half and put half back in plastic and return to refrigerator.

When Nematodes Come in a Sponge

We will need: source of water, tap water is fine, small container - bowl or small bucket, large bucket - 5 gallon or convenient size, sprinkling can or sprayer. Pour a small amount of water in the small container - about a quart will do. Cut open the bag. Place the sponge in the water and let it sit for 5 minutes. Rinse the plastic bag and pour wash water into the large container.

Shake, squeeze, and stretch the sponge, then pour the wash water into the large container. Pour more water into the small container. Place the sponge in the small container and let it sit for 5 more minutes. Shake, squeeze, and stretch again, then pour the wash water from into the large container one last time. Discard the sponge. Dilute nematodes in large container with enough water to easily cover the area to be treated.

Agitate Solution and keep Soil Moist

Agitate the water in the container while applying. Moisten the soil with water afterward to help the nematodes move into the soil. Avoid over-watering plants for one week after applying nematodes as they maybe washed out of the media.

Chapter 5

Microorganism

A microorganism or microbe is an organism that is microscopic (usually too small to be seen by the naked human eye). The study of microorganisms is called microbiology, a subject that began with Anton van Leeuwenhoek's discovery of microorganisms in 1675, using a microscope of his own design.

Microorganisms are very diverse. They include bacteria, fungi, archaea, and protists; microscopic plants (called green algae); and animals such as plankton, the planarian and the amoeba. Some also include viruses, but others consider these as non-living. Most microorganisms are unicellular (or single-celled), but this is not universal, since some multicellular organisms are microscopic, while some unicellular protists and bacteria, like *Thiomargarita namibiensis*, are macroscopic.

Fig. A Cluster of *Escherichia Coli* Bacteria Magnified 10,000 Times.

Microorganisms live in all parts of the biosphere where there is liquid water, including hot springs, on the ocean floor, high in the atmosphere and deep inside rocks within the Earth's crust. Microorganisms are critical to nutrient recycling in ecosystems as they act as decomposers.

As some microorganisms can fix nitrogen, they are a vital part of the nitrogen cycle, and recent studies indicate that airborne microbes may play a role in precipitation and weather.

Microbes are also exploited by people in biotechnology, both in traditional food and beverage preparation, and in modern technologies based on genetic engineering. However, pathogenic microbes are harmful, since they invade and grow within other organisms, causing diseases that kill millions of people, other animals, and plants.

EVOLUTION

Single-celled microorganisms were the first forms of life to develop on earth, approximately 3–4 billion years ago. Further evolution was slow, and for about 3 billion years in the Precambrian eon, all organisms were microscopic. So, for most of the history of life on Earth the only form of life were microorganisms. Bacteria, algae and fungi have been identified in amber that is 220 million years old, which shows that the morphology of microorganisms has changed little since the triassic period.

Most microorganisms can reproduce rapidly and microbes such as bacteria can also freely exchange genes by conjugation, transformation and transduction between widely-divergent species. This horizontal gene transfer, coupled with a high mutation rate and many other means of genetic variation, allows microorganisms to swiftly evolve to survive in new environments and respond to environmental stresses.

This rapid evolution is important in medicine, as it has led to the recent development of 'super-bugs' — pathogenic bacteria that are resistant to modern antibiotics.

PRE-MICROBIOLOGY

The possibility that microorganisms might exist was discussed for many centuries before their actual discovery in the 17th century. The first ideas about microorganisms were those of the Roman scholar Marcus Terentius Varro in a 1st century BC book titled *On Agriculture* in which he warns against locating a homestead near swamps:

This passage seems to indicate that the ancients were aware of the possibility that diseases could be spread by yet unseen organisms.

In *The Canon of Medicine* (1020), Abû Alî ibn Sînâ (Avicenna) stated that bodily secretion is contaminated by foul foreign earthly bodies before being infected. He also hypothesized that tuberculosis and other diseases might be contagious, *i.e.* that they were infectious diseases, and used quarantine to limit their spread.

When the Black Death bubonic plague reached al-Andalus in the

14th century, Ibn Khatima wrote that infectious diseases were caused by contagious "minute bodies" that enter the human body. Later, in 1546, Girolamo Fracastoro proposed that epidemic diseases were caused by transferable seedlike entities that could transmit infection by direct or indirect contact, or even without contact over long distances.

All these early claims about the existence of microorganisms were speculative in nature and not based on any data or science. Microorganisms were neither proven, observed, nor correctly and accurately described until the 17th century. The reason for this was that all these early inquiries lacked the most fundamental tool in order for microbiology and bacteriology to exist as a science, and that was the microscope.

MICROORGANISMS' DISCOVERY

Anton van Leeuwenhoek was the first person to observe microorganisms, using a microscope of his own design, thereby making him the first microbiologist. In doing so Leeuwenhoek would make one of the most important contributions to biology and open up the fields of microbiology and bacteriology. Prior to Leeuwenhoek's discovery of microorganisms in 1675, it had been a mystery as to why grapes could be turned into wine, milk into cheese, or why food would spoil.

Leeuwenhoek did not make the connection between these processes and microorganisms, but using a microscope, he did establish that there were forms of life that were not visible to the naked eye. Leeuwenhoek's discovery, along with subsequent observations by Lazzaro Spallanzani and Louis Pasteur, ended the long-held belief that life spontaneously appeared from non-living substances during the process of spoilage.

Lazzaro Spallanzani found that microorganisms could only settle in a broth if the broth was exposed to the air. He also found that boiling the broth would sterilise it and kill the microorganisms. Louis Pasteur expanded upon Spallanzani's findings by exposing boiled broths to the air, in vessels that contained a filter to prevent all particles from passing through to the growth medium, and also in vessels with no filter at all, with air being admitted via a curved tube that would not allow dust particles to come in contact with the broth.

By boiling the broth beforehand, Pasteur ensured that no microorganisms survived within the broths at the beginning of his experiment. Nothing grew in the broths in the course of Pasteur's experiment. This meant that the living organisms that grew in such broths came from outside, as spores on dust, rather than spontaneously generated within the broth. Thus, Pasteur dealt the death blow to the theory of spontaneous generation and supported germ theory.

In 1876, Robert Koch established that microbes can cause disease. He did this by finding that the blood of cattle who were infected with anthrax always had large numbers of *Bacillus anthracis.*

Koch also found that he could transmit anthrax from one animal to another by taking a small sample of blood from the infected animal and injecting it into a healthy one, causing the healthy animal to become sick.

He also found that he could grow the bacteria in a nutrient broth, inject it into a healthy animal, and cause illness. Based upon these experiments, he devised criteria for establishing a causal link between a microbe and a disease in what are now known as Koch's postulates. Though these postulates cannot be applied in all cases, they do retain historical importance in the development of scientific thought and can still be used today.

CLASSIFICATION AND STRUCTURE

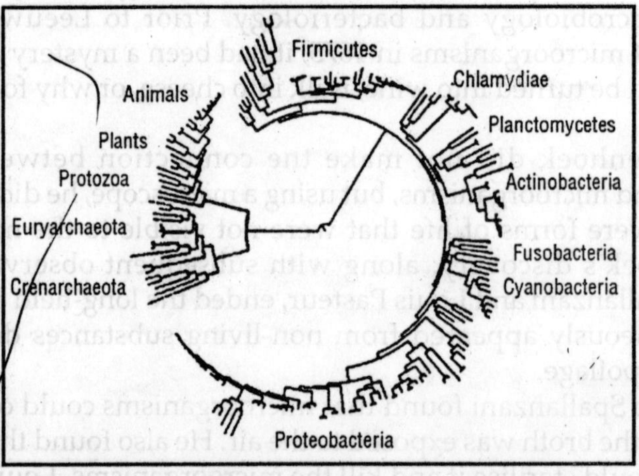

Fig. Common Ancestry of All Three Domains

Microorganisms can be found almost anywhere in the taxonomic organization of life on the planet.

Bacteria and archaea are almost always microscopic, while a number of eukaryotes are also microscopic, including most protists, some fungi, as well as some animals and plants.

Viruses are generally regarded as not living and therefore are not microbes, although the field of microbiology also encompasses the study of viruses.

Prokaryotes

Prokaryotes are organisms that lack a cell nucleus and the other organelles found in eukaryotes.

Prokaryotes are almost always unicellular, although some species such as myxobacteria can aggregate into complex structures as part of their life cycle. These organisms are divided into two groups, the archaea and the bacteria.

Bacteria

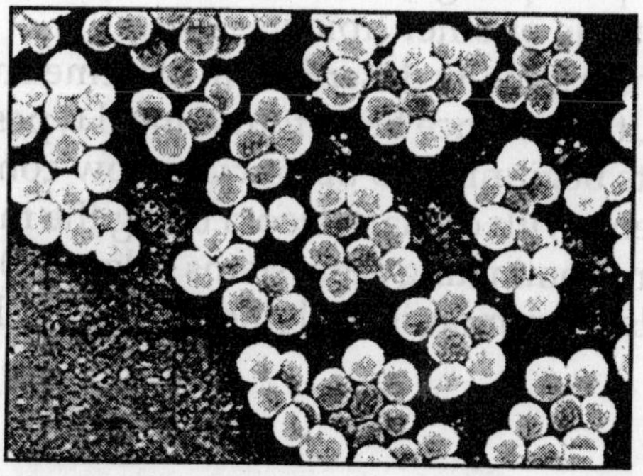

Fig. *Staphylococcus Aureus* Bacteria Magnified About 10,000x

Bacteria are the most diverse and abundant group of organisms on Earth. Bacteria inhabit practically all environments where some liquid water is available and the temperature is below +140 °C. They are found in sea water, soil, air, animals' gastrointestinal tracts, hot springs and even deep beneath the Earth's crust in rocks. Practically all surfaces which have not been specially sterilized are covered in bacteria. The number of bacteria in the world is estimated to be around five million trillion trillion, or 5×10^{30}.

Bacteria are practically all invisible to the naked eye, with a few extremely rare exceptions, such as *Thiomargarita namibiensis*. They are unicellular organisms and lack membrane-bound organelles. Their genome is usually a single loop of DNA, although they can also harbor small pieces of DNA called plasmids. These plasmids can be transferred between cells through bacterial conjugation.

Bacteria are surrounded by a cell wall, which provides strength and rigidity to their cells. They reproduce by binary fission or sometimes by budding, but do not undergo sexual reproduction. Some species form extraordinarily resilient spores, but for bacteria this is a mechanism for survival, not reproduction. Under optimal conditions bacteria can grow extremely rapidly and can double as quickly as every 10 minutes.

ARCHAEA

Archaea are also single-celled organisms that lack nuclei. In the past, the differences between bacteria and archaea were not recognised and archaea were classified with bacteria as part of the kingdom Monera. However, in 1990 the microbiologist Carl Woese proposed the three-domain system that divided living things into bacteria, archaea and eukaryotes. Archaea differ from bacteria in both their genetics and biochemistry. For example, while bacterial cell membranes are made from phosphoglycerides with ester bonds, archaean membranes are made of ether lipids.

Archaea were originally described in extreme environments, such as hot springs, but have since been found in all types of habitats. Only now are scientists beginning to appreciate how common archaea are in the environment, with crenarchaeota being the most common form of life in the ocean, dominating ecosystems below 150 m in depth. These organisms are also common in soil and play a vital role in ammonia oxidation.

EUKARYOTES

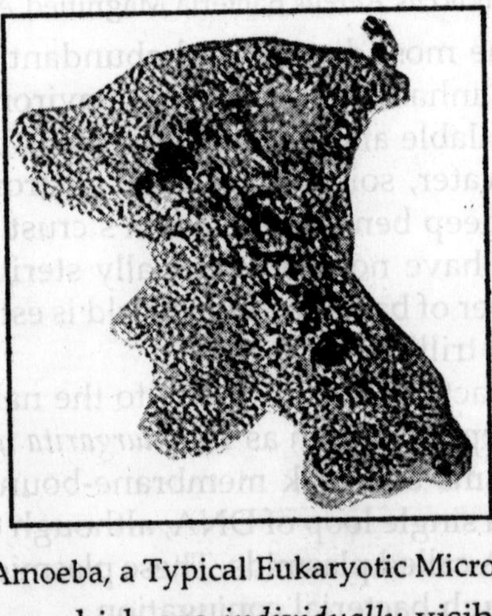

Fig. An Amoeba, a Typical Eukaryotic Microorganism

All living things which are *individually* visible to the naked eye are eukaryotes (with few exceptions; such as *Thiomargarita namibiensis*), including humans. However, a large number of eukaryotes are also microorganisms. Unlike bacteria and archaea, eukaryotes contain organelles such as the cell nucleus, the Golgi apparatus and mitochondria in their cells. The nucleus is an organelle which houses the DNA that makes up a cell's genome.

DNA itself is arranged in complex chromosomes. Mitochondria

are organelles vital in metabolism as they are the site of the citric acid cycle and oxidative phosphorylation. They evolved from symbiotic bacteria and retain a remnant genome. Like bacteria, plant cells have cell walls, and contain organelles such as chloroplasts in addition to the organelles in other eukaryotes. Chloroplasts produce energy from light by photosynthesis, and were also originally symbiotic bacteria.

Unicellular eukaryotes are those eukaryotic organisms that consist of a single cell throughout their life cycle. This qualification is significant since most multicellular eukaryotes consist of a single cell called a zygote at the beginning of their life cycles. Microbial eukaryotes can be either haploid or diploid, and some organisms have multiple cell nuclei. However, not all microorganisms are unicellular as some microscopic eukaryotes are made from multiple cells.

PROTISTS

Of eukaryotic groups, the protists are most commonly unicellular and microscopic. This is a highly diverse group of organisms that are not easy to classify. Several algae species are multicellular protists, and slime molds have unique life cycles that involve switching between unicellular, colonial, and multicellular forms. The number of species of protozoa is uncertain, since we may have identified only a small proportion of the diversity in this group of organisms.

Fig. A Microscopic Mite *Lorryia Formosa.*

ANIMALS

All animals are multicellular, but some are too small to be seen by the naked eye. Microscopic arthropods include dust mites and spider mites. Microscopic crustaceans include copepods and the cladocera, while many nematodes are too small to be seen with the naked eye.

Another particularly common group of microscopic animals are the rotifers, which are filter feeders that are usually found in fresh water. Micro-animals reproduce both sexually and asexually and may reach new habitats as eggs that survive harsh environments that would kill the adult animal. However, some simple animals, such as rotifers and nematodes, can dry out completely and remain dormant for long periods of time.

FUNGI

The fungi have several unicellular species, such as baker's yeast (*Saccharomyces cerevisiae*) and fission yeast (*Schizosaccharomyces pombe*). Some fungi, such as the pathogenic yeast *Candida albicans*, can undergo phenotypic switching and grow as single cells in some environments, and filamentous hyphae in others.

Fungi reproduce both asexually, by budding or binary fission, as well by producing spores, which are called conidia when produced asexually, or basidiospores when produced sexually.

PLANTS

The green algae are a large group of photosynthetic eukaryotes that include many microscopic organisms. Although some green algae are classified as protists, others such as charophyta are classified with embryophyte plants, which are the most familiar group of land plants. Algae can grow as single cells, or in long chains of cells.

The green algae include unicellular and colonial flagellates, usually but not always with two flagella per cell, as well as various colonial, coccoid, and filamentous forms. In the Charales, which are the algae most closely related to higher plants, cells differentiate into several distinct tissues within the organism. There are about 6000 species of green algae.

HABITATS AND ECOLOGY

Microorganisms are found in almost every habitat present in nature. Even in hostile environments such as the poles, deserts, geysers, rocks, and the deep sea. Some types of microorganisms have adapted to the extreme conditions and sustained colonies; these organisms are known as extremophiles.

Extremophiles have been isolated from rocks as much as 7 kilometres below the earth's surface, and it has been suggested that the amount of living organisms below the earth's surface may be comparable with the amount of life on or above the surface. Extremophiles have been known to survive for a prolonged time in a vacuum, and can be highly resistant to radiation, which may even allow them to survive in space.

Many types of microorganisms have intimate symbiotic relationships with other larger organisms; some of which are mutually beneficial (mutualism), while others can be damaging to the host organism (parasitism). If microorganisms can cause disease in a host they are known as pathogens.

EXTREMOPHILES

Extremophiles are microorganisms which have adapted so that they can survive and even thrive in conditions that are normally fatal to most lifeforms. For example, some species have been found in the following extreme environments:

- Temperature: as high as 130 °C (266 °F), as low as "17 °C (1.4 °F).
- Acidity/alkalinity: less than pH 0, up to pH 11.5.
- Salinity: up to saturation.
- Pressure: up to 1,000-2,000 atm, down to 0 atm (e.g. vacuum of space).
- Radiation: up to 5kGy.

Extremophiles are significant in different ways. They extend terrestrial life into much of the Earth's hydrosphere, crust and atmosphere, their specific evolutionary adaptation mechanisms to their extreme environment can be exploited in bio-technology, and their very existence under such extreme conditions increases the potential for extraterrestrial life.

SOIL MICROBES

The nitrogen cycle in soils depends on the fixation of atmospheric nitrogen. One way this can occur is in the nodules in the roots of legumes that contain symbiotic bacteria of the genera *Rhizobium, Mesorhizobium, Sinorhizobium, Bradyrhizobium*, and *Azorhizobium*.

Importance

Microorganisms are vital to humans and the environment, as they participate in the Earth's element cycles such as the carbon cycle and nitrogen cycle, as well as fulfilling other vital roles in virtually all ecosystems, such as recycling other organisms' dead remains and waste products through decomposition.

Microbes also have an important place in most higher-order multicellular organisms as symbionts. Many blame the failure of Biosphere 2 on an improper balance of microbes.

Use in Food

Microorganisms are used in brewing, winemaking, baking, pickling and other food-making processes. They are also used to control the fermentation process in the production of cultured dairy products such as yogurt and cheese. The cultures also provide flavour and aroma, and inhibit undesirable organisms.

Use in Water Treatment

Specially-cultured microbes are used in the biological treatment of sewage and industrial waste effluent, a process known as bioaugmentation.

Use in Energy

Microbes are used in fermentation to produce ethanol, and in biogas reactors to produce methane. Scientists are researching the use of algae to produce liquid fuels, and bacteria to convert various forms of agricultural and urban waste into usable fuels.

Use in Science

Microbes are also essential tools in biotechnology, biochemistry, genetics, and molecular biology.

The yeasts (*Saccharomyces cerevisiae*) and fission yeast (*Schizosaccharomyces pombe*) are important model organisms in science, since they are simple eukaryotes that can be grown rapidly in large numbers and are easily manipulated.

They are particularly valuable in genetics, genomics and proteomics. Microbes can be harnessed for uses such as creating steroids and treating skin diseases. Scientists are also considering using microbes for living fuel cells, and as a solution for pollution.

Use in Warfare

In the Middle Ages, diseased corpses were thrown into castles during sieges using catapults or other siege engines. Individuals near the corpses were exposed to the deadly pathogen and were likely to spread that pathogen to others.

IMPORTANCE IN HUMAN HEALTH

Human Digestion

Microorganisms can form an endosymbiotic relationship with other, larger organisms. For example, the bacteria that live within the human digestive system contribute to gut immunity, synthesise vitamins such as folic acid and biotin, and ferment complex indigestible carbohydrates.

Diseases and Immunology

Microorganisms are the cause of many infectious diseases. The organisms involved include pathogenic bacteria, causing diseases such as plague, tuberculosis and anthrax; protozoa, causing diseases such as malaria, sleeping sickness and toxoplasmosis; and also fungi causing diseases such as ringworm, candidiasis or histoplasmosis.

However, other diseases such as influenza, yellow fever or AIDS are caused by pathogenic viruses, which are not usually classified as living organisms and are not therefore microorganisms by the strict definition. As of 2007, no clear examples of archaean pathogens are known, although a relationship has been proposed between the presence of some methanogens and human periodontal disease.

IMPORTANCE IN ECOLOGY

Microbes are critical to the processes of decomposition required to cycle nitrogen and other elements back to the natural world.

HYGIENE

Hygiene is the avoidance of infection or food spoiling by eliminating microorganisms from the surroundings. As microorganisms, particularly bacteria, are found practically everywhere, this means in most cases the reduction of harmful microorganisms to acceptable levels. However, in some cases it is required that an object or substance be completely sterile, i.e. devoid of all living entities and viruses. A good example of this is a hypodermic needle.

In food preparation microorganisms are reduced by preservation methods (such as the addition of vinegar), clean utensils used in preparation, short storage periods or by cool temperatures. If complete sterility is needed, the two most common methods are irradiation and the use of an autoclave, which resembles a pressure cooker.

There are several methods for investigating the level of hygiene in a sample of food, drinking water, equipment etc. Water samples can be filtrated through an extremely fine filter. This filter is then placed in a nutrient medium. Microorganisms on the filter then grow to form a visible colony.

Harmful microorganisms can be detected in food by placing a sample in a nutrient broth designed to enrich the organisms in question. Various methods, such as selective media or PCR, can then be used for detection. The hygiene of hard surfaces, such as cooking pots, can be tested by touching them with a solid piece of nutrient medium and then allowing the microorganisms to grow on it.

There are no conditions where all microorganisms would grow, and therefore often several different methods are needed. For example, a food sample might be analyzed on three different nutrient mediums designed to indicate the presence of "total" bacteria (conditions where many, but not all, bacteria grow), molds (conditions where the growth of bacteria is prevented by e.g. antibiotics) and coliform bacteria (these indicate a sewage contamination).

METALLURGY

Metallurgy, science and technology of metals, including the extraction of metals from ores, the preparation of metals for use, and the study of the relationship between structures and properties of metals.

Metallurgical processes consist of two operations: concentration, separating a metal or metallic compound from the useless waste rock material, or gangue, which accompanies it in the ore; and refining, producing the metal in a pure or nearly pure state suitable for use. Three types of processes are employed both for concentration and refining: mechanical, chemical, and electrical. In most cases a combination of these methods is used.

One of the simplest methods of mechanical separation is gravity separation. This process is based on the difference in specific gravity between native metals and metallic minerals, and the other rock materials with which they are mixed. When crushed ore or ore concentrates are suspended either in water or an air blast, the heavier metal or metallic mineral particles fall to the bottom of the processing chamber, and the lighter gangue is blown or washed away.

The prospector's technique of panning gold from gold-bearing sand, for example, is a small-scale gravity-separation process. Similarly, by virtue of its higher specific gravity, magnetite, a mineral of iron, may be separated from the gangue rock in which it occurs. Flotation is the most important present-day method of mechanical concentration. In its simplest form, flotation is a modified gravity process in which finely ground ore is mixed, usually with a liquid.

The metal or metallic mineral floats while the gangue sinks, although the reverse is true in some instances. In most modern flotation processes, the floating of either the metal or gangue is aided by an oil or other surface-active agent. By this means, comparatively heavy substances can be made to float on water. In one typical process, a finely ground ore containing copper sulfide is mixed with water, to which small amounts of oil, acid, or other so-called flotation reagents are added. When air is blown through this mixture, a froth is formed on the surface that has the property of mixing with the sulfide but not with the gangue.

The latter material settles, and the sulfide is collected from the froth. Use of the flotation process has made possible the exploitation of many ore deposits of low concentration, and even of the wastes from processing plants that used less efficient techniques. In some cases, by means of differential flotation, different minerals can be concentrated from one complex ore in a single process.

Ores, such as magnetite, that have marked magnetic properties are concentrated by means of electromagnets that attract the metal but do not attract the gangue. Electrostatic separation employs an electric field to separate minerals of different electrical properties by exploiting the attraction between unlike charges and the repulsion between like charges.

Chemical separation methods are, in general, the most important from the economic point of view. In present-day practice chemical separation often is used as a second stage after mechanical concentration. A greater tonnage of refined metal is obtained by smelting than by any other process. In smelting, the ore, or the concentrate from a mechanical separation process, is heated with a reducing agent and a flux to a high temperature.

The reducing agent combines with the oxygen in a metallic oxide, leaving pure metal; and the flux combines with the gangue to form a slag that is liquid at the smelting temperature and thus can be skimmed off or poured away from the metal. The production of pig iron in blast furnaces is an example of smelting, and the process is also used to extract copper, lead, nickel, and many other metals from their ores.

Amalgamation is a metallurgical process that utilizes mercury to dissolve silver or gold to form an amalgam. This process has been largely supplanted by the cyanide process, in which gold or silver is dissolved in solutions of sodium or potassium cyanide. Various types of aqueous solutions are employed in different leaching, or percolating, processes to dissolve metals from ores.

Metallic carbonates and sulfides are treated by roasting, heating to a temperature below the melting point of the metal. In the case of carbonates, carbon dioxide is driven off in the process, leaving a metallic oxide. When sulfides are roasted, the sulfur combines with the oxygen of the air to form gaseous sulfur dioxide, leaving metallic oxides, which are subsequently reduced by smelting.

Agglomeration of ore fines (fine particles) is accomplished by sintering or pelletizing. In the sintering process, fuel, water, air, and heat are used to fuse the ore fine into a porous mass. In pelletizing, moistened fine is formed into small pellets in the presence of limestone flux and then fired.

A number of other processes, of which pyrometallurgy (high-temperature metallurgy) and distillation are the most important, are employed in further refinement stages of a variety of metals. In the process of electrolysis the metal is deposited at the cathode from aqueous solutions or in an electrolytic furnace.

Copper, nickel, zinc, silver, and gold are several examples of metals that are refined by deposition from aqueous solutions. Aluminum, barium, calcium, magnesium, beryllium, potassium, and sodium are metals that are processed in electrolytic furnaces.

GENERAL SEQUENCE OF OPERATIONS

Occasionally, but rarely, metallic ores occur as practically pure compact masses, from which the accompanying matrix or "gangue" can be detached by hand and hammer. In most cases the "ore", as it comes out of the mine or quarry, is simply a mixture of ore proper and gangue, in which the latter not unfrequently predominates.

Hence it is generally necessary to purify the ore before the liberation of the metal is attempted. Most metallic ores are specifically heavier than the accompanying impurities and their purification is generally effected by reducing the crude ore to a fine enough powder to detach the metallic from the earthy part, and then washing away the latter by a current of water, as far as possible.

The majority of ores being chemical compounds, the extraction of their metals demands chemical treatment. The chemical operations involved may be classified as follows: I. *Fiery Operations. - The* ore, generally with some "flux," is exposed to the action of fire. The fire in most cases has a chemical, in addition to its physical, function. Moreover the furnace is designed so as to facilitate the action of the heat and furnace gases in the desired direction.

It is intended either to burn away certain components of the ore - in which case it must be so regulated as to contain a sufficient excess of unburned oxygen; or it is meant to deoxidize ("reduce") the ore,when the draught must be restricted so as to keep the ore constantly wrapped up in combustible flame gases (carbon monoxide, hydrogen, marsh-gas, &c.).

The majority of the chemical operations of metallurgy fall into this category, and in these processes other metal-reducing agents than those naturally contained in the fire (or blast) are only exceptionally employed.

AMALGAMATION

The ore by itself (if it is a reguline one), or with certain reagents (if it is not), is worked up with mercury so that the metal is obtained as an amalgam, which can be separated mechanically from the dross. The purified amalgam is distilled, when the mercury is recovered as a distillate while the metal remains.

WET PROCESSES

Strictly speaking, certain amalgamation methods fall under this head; but, in its ordinary acceptance, the term refers to processes in which the metal is extracted either from the natural ore, or from the ore after roasting or other preliminary treatment, by an acid or salt solution, and from this solution precipitated - generally in the reguline form - by some suitable reagent.

Few methods of metal extraction at once yield a pure product. What as a rule is obtained is a more or less impure metal, which requires to be "refined" to become fit for the market.

CHEMICAL OPERATIONS

Amalgamation and wet-way processes have limited applications, being practically confined to copper, gold and silver. We therefore here confine ourselves, in the main, to pyro-chemical operations.

The method to be adopted for the extraction of a metal from its ore is determined chiefly, though not entirely, by the nature of the non-metallic component with which the metal is combined. The simplest case is that of the reguline ores where there is no nonmetallic element. The important case is that of gold.

OXIDES, HYDRATES, CARBONATES AND SILICATES

All iron and tin ores proper fall under this heading, which, besides, comprises certain ores of copper, of lead and of zinc. The first step consists in subjecting the crude ore to a roasting or calcining process, the object of which is to remove the water and carbon dioxide, and burn away, to some extent at least, the sulphur, arsenic or organic matter.

The residue consists of an impure oxide of the respective metal, which in all cases is reduced by treatment with fuel at a high temperature. Should the metal be present as a silicate, lime must be added in the smelting to remove the silica and liberate the oxide.

The temperature required for the reduction of zinc lies above the boiling point of the metal; hence the mixture of ore and reducing agent (charcoal is generally used) must be heated in a retort combined with condensing apparatus. In all the other cases the reduction is effected in the fire itself, a tower-shaped blast furnace being preferably used.

The furnace is charged with alternate layers of fuel and ore and the whole kindled from below. The metallic oxide, partly by the direct action of the carbon with which it is in contact, but principally by that of the carbon monoxide produced in the lower strata from the oxygen of the blast and the hot carbon there, is reduced to the metallic state; the metal fuses and runs down, with the slag, to the bottom of the furnace, whence both are withdrawn by opening plug-holes.

SULPHIDES

Iron, copper, lead, zinc, mercury, silver and antimony very frequently present themselves in this state of combination, as components of a family of ores which may be divided into two sections:

- Such as substantially consist of simple sulphides, as iron pyrites (FeS_2), galena (PbS), zinc blende (ZnS), cinnabar (HgS); and
- Complex sulphides, such as the various kinds of sulphureous copper ores (all substantially compounds or mixtures of sulphides of copper and iron); bournonite, a complex sulphide of lead, antimony and copper; rothgiltigerz, sulphide of silver, antimony and arsenic; fahlerz, sulphides of arsenic and antimony, combined with sulphides of copper, silver, iron, zinc, mercury, silver; and mixtures of these and other sulphides with one another.

In treating a sulphureous ore, the first step as a rule is to subject it to oxidation by roasting it in a reverberatory or other furnace, which leads to the burning away of at least part of the arsenic and part of the sulphur.

The effect on the several individual metallic sulphides (supposing only one of these to be present) is as follows:

- Those of silver (Ag_2S) and mercury (HgS) yield sulphur dioxide gas and metal; in the case of silver, sulphate is formed at low temperatures. Metallic mercury, in the circumstances, goes off as a vapour, which is collected and condensed; silver remains as a regulus, but pure sulphide of silver is hardly ever worked.
- Sulphides of iron and zinc yield the oxides Fe_2O_3 and ZnO as final products, some basic sulphate being formed at the

earlier stages, especially in the case of zinc. The oxides can be reduced by carbon.

- The sulphides of lead and copper yield, the former a mixture of oxide and normal sulphate, the latter one of oxide and basic sulphate. Sulphate of lead is stable at a red heat; sulphate of copper breaks up into oxide, sulphur dioxide and oxygen. In practice, neither oxidation process is ever pushed to the end; it is stopped as soon as the mixture of roasting-product and unchanged sulphide contains oxygen and sulphur in the ratio of 0 2: S. The access of air is then stopped and the whole heated to a higher temperature, when the whole of the sulphur and oxygen is eliminated. This method is largely utilized in the smelting of lead from galena and of copper from copper pyrites.
- Sulphide of antimony, when roasted in air, is converted into a kind of alloy of sulphide and oxide; the same holds for iron, only its oxysulphide is quite readily converted into the pure oxide Fe_2O_3 by further roasting. Oxysulphide of antimony, by suitable processes can be reduced to metal, but these processes are rarely used, because the same end is far more easily obtained by "precipitation," *i.e.* withdrawing the sulphur by fusion with metallic iron, forming metallic antimony and sulphide of iron. Both products fuse, but readily part, because fused antimony is far heavier than fused sulphide of iron. A precisely similar method is used occasionally for the reduction of lead from galena. Sulphide of lead, when fused together with metallic iron in the proportion of 2Fe: I PbS yields a regulus (= I Pb) and a "mat" Fe_2S, which, however, on cooling, decomposes into the ordinary sulphide FeS, and finely divided iron. What we have been explaining are special cases of a more general metallurgic proposition: Any one of the metals, copper, iron, tin, zinc, lead, silver, antimony, arsenic, in general, is capable of desulphurizing (at least partially) any of the others that follows it in the series just given, and it does so the more readily and completely the greater the number of intervening terms. Hence, supposing a complete mixture of these metals to be melted down under circumstances admitting of only a partial sulphuration of the whole, the copper has the best chance of passing into the "mat," while the arsenic is the first to be eliminated as such, or, in the presence of oxidants, as oxide.

ARSENIDES

Although arsenides are amongst the commonest impurities of ores generally, ores consisting essentially of arsenides are comparatively rare.

The most important are certain double arsenides of cobalt and nickel, which in practice are always contaminated with the arsenides or other compounds of foreign metals, such as iron, manganese, &c. The general mode of working these ores is as follows.

The ore is first roasted by itself, when a part of the arsenic goes off as such and as oxide, while a complex of lower arsenides remains. This residue is now subjected to careful oxidizing fusion in the presence of some solvent for metallic bases. The effect is that the several metals are oxidized away and pass into the slag (as silicates) in the following order - manganese, iron, cobalt, nickel; and at any stage the as yet unoxidized residue of arsenide assumes the form of a fused regulus, which sinks down through the slag as a "speis."

By stopping the process at the right moment, we can produce a speis which contains only cobalt and nickel, and if at this stage also the flux is renewed we can further produce a speis which contains only nickel and a slag which substantially is one of cobalt only.

Minor Reagents

Besides the oxidizing and reducing agents present in the fire, and the "fluxes" added for the production of slags, various minor reagents may be noticed. Metallic iron as a desulphurizer has already been referred to. Oxide of lead, PbO (litharge), is largely used as an oxidizing agent. At a red heat, when it melts, it readily attacks all metals, except silver and gold, the general result being the formation of a mixed oxide and of a mixed regulus, a distribution, in other words, of both the lead and the metal acted on between slag and regulus.

More important is its action on metallic sulphides, which, in general, results in the formation of three things besides sulphur dioxide, viz. a mixed oxide slag including the excess of litharge, a regulus of lead (which may include bismuth and other more readily reducible metals), and, if the litharge is not sufficient for a complete oxidation, a "mat" comprising the more readily sulphurizable metals. Oxide of lead, being a most powerful solvent for metallic oxides generally, is also largely used for the separation of silver or gold from base metallic oxides.

Metallic lead is to metals generally what oxide of lead is to metallic oxides. It accordingly is available as a solvent for taking up small particles of metal diffused throughout a mass of slag, and uniting them into one regulus. This leads us to the process of "cupellation," which serves for the extraction of gold and silver from their alloys with base metals.

FLUXES

All ores are contaminated with more or less gangue, which in general consists of infusible matter, and if left unheeded in the reduction of the metallic part of the ore would retain more or less of the metal disseminated through it, or at best foul the furnace. To avoid this, the ore as it goes into the furnace is mixed with "fluxes" so selected as to convert the gangue into a fusible "slag," which readily runs down through the fuel with the regulus and separates from the latter.

The quality and proportion of flux should, if possible, be so chosen that the formation of the slag sets in only after the metal has been reduced and molten; or else part of the basic oxide of the metal to be extracted may be dissolved by the slag and its reduction thus be prevented or retarded.

Slags are not a necessary evil; if an ore were free from gangue we should add gangue and flux from without to produce a slag, because one of its functions is to form a layer on the regulus which protects it against the further action of the blast or furnace gases. Fluxes may be arranged under the three heads of

- Fluor-spar
- Basic fluxes and
- Acid fluxes

Fluor-spar fuses up at a red heat with silica, sulphates of calcium and barium, and a few other infusible substances into homogeneous masses. It shows little tendency to dissolve basic oxides, such as lime. One part of fluor-spar liquefies about half a part of silica, four parts of calcium sulphate and one and a half parts of barium sulphate. Upon these facts its extremely wide application in metallurgy is founded. Carbonate of soda (or potash) is the most powerful basic flux.

It dissolves silica and all silicates into fusible glasses. On the other hand, borax may be taken as a type for the acid fluxes. At a red heat, when it forms a viscid fluid, it readily dissolves P all basic oxides into fusible complex borates.

Now the gangue of an ore in general consists either of some basic material such as carbonate of lime (or magnesia), ferric oxide, alumina or of silica (quartz) or some more or less acid silicate, or else of a mixture of the two classes of bodies. So any kind of gangue might be liquefied by means of borax or by means of alkaline carbonate; but neither of the two is used otherwise than for assaying; what the metal-smelter does is to add to a basic gangue the proportion of silica, and to an acid ore the proportion of lime, or, indirectly, of ferrous or perhaps manganous oxide, which it may need for the formation of a slag of the proper qualities.

The slag must possess the proper degree of saturation. In other words, taking SLO_2+ nMeO (where MeO means an equivalent of base) as a formula for the potential slag, n must have the proper value. If n is too small, *i.e.* if the slag is too acid, it may dissolve part of the metal to be recovered; if *n* is too great, *i.e.* the slag too basic, it may refuse to dissolve, for instance, the ferrous oxide which is meant to go into it, and this oxide will then be reduced, and its metal (iron in our example) contaminate the regulus.

In reference to the problem under discussion, it is worth noting that oxides of lead and copper are more readily reduced to metals than oxide of iron Fe_2O_3 is to FeO, the latter more readily to FeO than FeO itself to metal, and FeO more readily to metal than manganous oxide is. Oxide of calcium (lime) is not reducible at all. The order of basicity in the oxides (their readiness to go into the slag) is precisely the reverse.

Most slags being complex silicates, it is a most important problem of scientific metallurgy to determine the relations in this class of bodies between chemical composition on the one hand and fusibility and solvent power for certain oxides (CaO, FeO, Fe_2O_3, AL_2O_3, SLO_2) on the other. Their general composition may be expressed as

n(MO- FxSiO_2)+m[(fe or al) OxSLO_2] (M =Ca, Mg, Fe, K_2 fe = 3Fe, al =1A1.)

CONCENTRATION OF THE ORE

When an ore has a low percentage of the desired metal, a method of physical concentration must be used before the extraction process begins. In one such method, the ore is crushed and placed in a machine where, by shaking, the heavier particles containing the metal are separated from the lighter rock particles by gravity. Another method is the flotation process, used commonly for copper sulfide ores.

In certain cases (as when gold, silver, or occasionally copper occur "free," i.e., uncombined chemically in sand or rock), mechanical or ore dressing methods alone are sufficient to obtain relatively pure metal. Waste material is washed away or separated by screening and gravity; the concentrated ore is then treated by various chemical processes.

SEPARATION OF THE METAL

Processes for separating the metal from the impurities it is found with or the other elements with which it is combined depend upon the chemical nature of the ore to be treated and upon the properties of the metal to be extracted. Gold and silver are often removed from the impurities associated with them by treatment with mercury, in which they are soluble. Another method for the separation of gold and silver is the so-called cyanide process.

The Parkes process, which is based on silver being soluble in molten zinc while lead is not, is used to free silver from lead ores. Since almost all the metals are found combined with other elements in nature, chemical reactions are required to set them free. These chemical processes are classified as pyrometallurgy, electrometallurgy, and hydrometallurgy.

Pyrometallurgy, or the use of heat for the treatment of an ore, includes smelting and roasting. If the ore is an oxide, it is heated with a reducing agent, such as carbon in the form of coke or coal; the oxygen of the ore combines with the carbon and is removed in carbon dioxide, a gas. The waste material in the ore is called gangue; it is removed by means of a substance called a flux which, when heated, combines with it to form a molten mass called slag. Being lighter than the metal, the slag floats on it and can be skimmed or drawn off.

The flux used depends upon the chemical nature of the ore; limestone is usually employed with a siliceous gangue. A sulfide ore is commonly roasted, i.e., heated in air. The metal of the ore combines with oxygen of the air to form an oxide, and the sulfur of the ore also combines with oxygen to form sulfur dioxide, which, being a gas, passes off. The metallic oxide is then treated with a reducing agent. When a carbonate ore is heated, the oxide of the metal is formed, and carbon dioxide is given off; the oxide is then reduced.

Electrometallurgy includes the preparation of certain active metals, such as aluminum, calcium, barium, magnesium, potassium, and sodium, by electrolysis: a fused compound of the metal, commonly the chloride, is subjected to an electric current, the metal collecting at the cathode.

Hydrometallurgy, sometimes called leaching, involves the selective dissolution of metals from their ores. For example, certain copper oxide and carbonate ores are treated with dilute sulfuric acid, forming water-soluble copper sulfate. The metal is recovered by electrolysis of the solution. If the metal obtained from the ore still contains impurities, special refining processes are required.

ORES AND MINERALS

Metals occur in nature sometimes in free or native state, but most of these occur in nature in the form of chemical combination, i.e., in the form of their stable compounds which are associated with gangue or matrix. The earthy impurities such as sand, clay, rocks etc. associated with ore are collectively known as gangue or matrix. Thus a large number of metals in nature occur in the combined form along with other elements, but some metals, such as Ag, Au, Pt etc. occur in the native form (metallic state) in some regions. Ag occurs in native (or free) as well as in the form of compounds.

The natural material in which the metal or their compounds occur in the earth is known as mineral. Thus a mineral is a naturally occurring material present in earth's crust which contains metal in the native (or free state) or in combined state. A mineral may be a single compound or a complex mixture of various compounds. When a mineral contains sufficient amount of a metal in combined state, from which it can be readily and profitably separated on commercial scale, then the mineral is said to be an ore of the metal.

Thus all ores are minerals, but all minerals are not ores. A mineral from which a metal can be profitably extracted is called an ore. For example, clay ($Al_2O_3\ 2\ SiO_2\ 2\ H_2O$) and bauxite ($Al_2\ O_3\ 2H_2O$) are two minerals of aluminium, but aluminium, can be profitably extracted only from bauxite and not from the clay. Hence bauxite is an ore of aluminium, while clay is a mineral.

The biggest source of metal is the earth's crust and the process of taking out the ores from the earth crust is called mining. In the combined state ores are generally found in the form of oxides, sulphides, carbonates, sulphates, chlorides and silicates.

Sources from Earth, Living Systems and in Sea Sources from Earth

Nearly 80 elements are obtained from mineral deposits on or beneath the surface of the earth.

- Metals which are sufficiently unreactive to occur uncombined (ie in elementary form) are present in group 10 and 11 of the 2nd and 3rd transition series (e.g Pt, Au, Ag; free Ag is also found in nature).
- Metalloids (e.g Ge, As, Sb) and neighbouring metals, all of which have relatively large ionization energies, generally occur as sulphides.
- The more strongly metallic elements that form positive ions readily are found as oxides (transition metals), carbonates (group 2 metals) or chlorides (group 1 metals).
- Three noble metals (Cu, Ag, Au), Hg and six platinum metals (Ru, Os, Rh, Ir, Pd and Pt) occur in nature in free state. All other metals are found in combined state in the nature.

Table. Classification of Ores

Ore	Ore or Mineral	Composition	Metal Present
Oxide ores	Bauxite	$Al_2O_3.2H_2O$	Al
	Cuprite	Cu_2O	Cu
	Haematite	Fe_2O_3	Fe
	Zincite	ZnO	Zn
	Tinstone or Casseterite	SnO_2	Sn
	Pyrolusite	MnO_2	Mn
	Pitch Blende	U_3O_8	U
Sulphide ores	Copper Pyrites	Cu_2S, Fe_2S_3 or $Cu_2S . FeS_2$	Cu
	Copper Glance	Cu_2S	Cu
	Zinc Blende	ZnS	Zn
	Cinnabar	HgS	Hg
	Galena	PbS	Pb
	Argentite or Silver Glance	Ag_2S	Ag
Carbonate ores	Magnesite	$MgCO_3$	Mg
	Dolomite	$CaCO_3.MgCO_3$	Mg
	Calamine	$ZnCO_3$	Zn
	Malachite	$CuCO_3.Cu(OH)_2$	Cu
	Limestone	$CaCO_3$	Ca
Halide ores	Rock Salt	$NaCl$	Na
	Carnallite	$KCl.MgCl_2.6H_2O$	Mg
	Horn Silver	$AgCl$	Ag
	Sylvine	KCl	K
	Cryolite	$3 NaF.AlF_3$ or Na_3AlF_6	Al
Sulphate ores	Epsom Salt	$MgSO_4.7H_2O$	Mg
	Gypsum	$CaSO_4.2H_2O$	Ca
	Barytes	$BaSO_4$	Ba
	Anglesite	$PbSO_4$	Pb
Silicate ores	Asbestos	$CaSiO_3.3MgSiO_3$	Mg
	Felspar	$K_2O.Al_2O_3.6SiO_2$ or $KAlSi_3O_8$	Al
	Mica	$K_2O.3Al_2O_3.6SiO_22H_2O$	Al
Phosphate ores	Phosphorite	$Ca_3(PO_4)_2$	P

- The composition of earth's crust is : O (49.1%), Si (26%), Al(7.5%), Fe(4.2%), Ca (3.2%), Na (2.4%), Mg (2.3%) and H (1.0%) by weight.
- In combined state metals are found as (a) Oxides- Mg, Cu, Zn, Al, Mn, Fe, etc. (b)Carbonates-Na, Cu, Mg, Ca, Ba, Zn, Fe etc.
- Phosphates- Ca, Pb, Fe etc (d) Silicates- Li, Cu, Zn, Al, Mn, Ni, etc and (e) Sulphates- Fe, Cu, Pb, Hg etc.

Source from Sea

Four elements such as Na, Mg, Cl_2 and Br_2 can be extracted from the oceans or salt brines, where they are present as monoatomic ions (Na^+, Mg^{2+}, Cl^-, Br^-).

Source from Living System

The descending mass abundance of elements. About 30 per cent of enzymes have a metal atom at the active site. A large number of biomolecules contain metal ions ; many of these molecules are proteins. In addition metal ions in the form of crystalline minerals or amorphous solids are important as structural materials in many organisms.

Table. Descending Mass Abundance of the Elements

Earth Crust	Oceans	Human
O	O	H
Si	H	O
Al	Cl	C
Fe	Na	N
Ca	Mg	Na
Mg	S	K
Na	Ca	Ca
K	K	Mg
Ti	C	P
H	Br	S
P	B	Cl

PURIFICATION OF ORES

The ore is generally associated with rock impurities like clay, sand etc. called 'gangue or matrix'. The purification of ore means removal of gangue from the powdered ore. This process is otherwise called concentration of the ore or ore dressing.

Thus, the percentage of the ore in the concentrated ore is higher than that in the original ore. The purification or concentration of ore can be brought about in the following ways, depending upon the type of ore.

Gravity Separation Process or Hydraulic Washing

This method is especially suitable for heavy 'oxide' ores like haematite, tinstone, etc. In this, the powdered ore is placed on a sloping floor (or platform) and washed by directing on it a strong current of water. The lighter sandy, and earthy impurities are washed away; while the heavier ore particles are left behind.

Froth Flotation Process

This method is especially suitable for sulphide ores like zinc blende (ZnS), and copper pyrites ($CuFeS_2$). This process is based on the fact that the sulphide ore particles are only moistened by oil; while those of oxide, and gangue particles are moistened only by water. In this process, the powdered ore is mixed with water and a little pine oil (a foaming agent) and the whole mixture is then stirred vigorously by blowing compressed air. The oil forms a foam (or froth) with air. The ore particles stick to the froth, which rises to the surface; while the rocky, and earthy impurities (gangue) are left in water. The froth is skimmed off, collected, and allowed to subside to get concentrated ore.

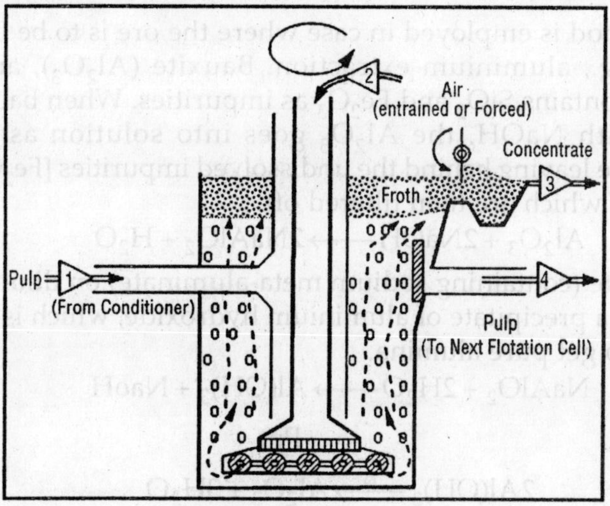

Fig. Froth Flotation Process

Electromagnetic Separation Process

This method is meant for separating magnetic impurities from nonmagnetic ore particles, e.g., tinstone (a tin ore) in which tinstone is non- magnetic; while impurities iron, manganese and tungstates are magnetic.

The powdered ore (containing the associated magnetic impurities) is made to fall (from a hopper) on a belt moving over electromagnetic roller. The magnetic impurities fall from the belt in a heap near the magnet, due to attraction; while the non-magnetic concentrated ore falls in separate heap, away from the magnet, due to the influence of centrifugal force.

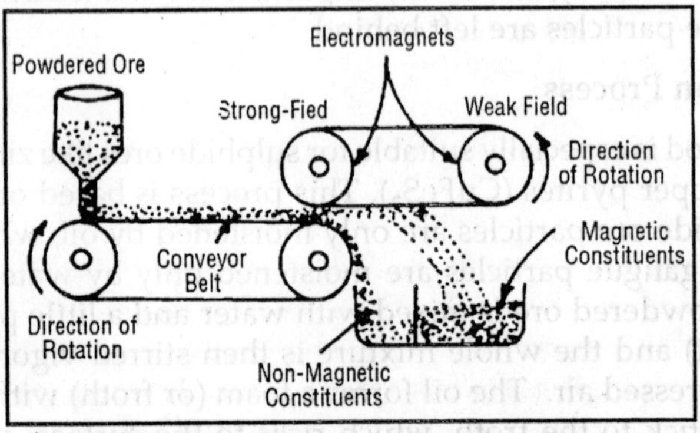

Fig. Electromagnetic Separation Method

Chemical Method

This method is employed in case where the ore is to be in a very pureform, e.g., aluminium extraction. Bauxite (Al_2O_3), an ore of aluminium, contains SiO_2 and Fe_2O_3 as impurities. When bauxite ore is treated with NaOH, the Al_2O_3 goes into solution as sodium metaaluminate leaving behind the undissolved impurities [Fe_2O_3, SiO_2, $Fe(OH)_3$,etc.], which are then filtered off.

$$Al_2O_3 + 2NaOH \longrightarrow 2NaAlO_2 + H_2O$$

The filtrate (containing sodium meta-aluminate) on dilution, and stirring gives a precipitate of aluminium hydroxide, which is filtered, and ignited to get pure alumina.

$$NaAlO_2 + 2H_2O \longrightarrow \underset{\text{Ppt}}{Al(OH)_3} + NaoH$$

$$2Al(OH)_3 \xrightarrow{\Delta} \underset{\text{Pure}}{Al_2O_3} + 3H_2O$$

METALLURGICAL PROCESSES

Metallurgy is a branch of chemistry which deals with,

- Extraction of metals from ores.
- Refining of crude metal.
- Producing alloys and the study of their constitution, structure and properties.
- The relationship of physical and mechanical treatment of metals to alloys.

The extraction of metals cannot be carried out by any universal method because extraction of each metal requires different methods of extraction. This depends upon the nature and preparation of metals. In general, noble metals such as Au, Ag, etc are usually extracted by electrolysis of their chlorides, oxides or hydroxides. Heavy metals, e.g. Cu, Zn, Fe, Pb, Sn, etc., are extracted by making use of roasting and smelting methods.

Roasting Oxidation

Roasting is one of the oxidation method where ore is converted into metal oxide. In the process of roasting, the ore either alone or with the addition of suitable material, is subjected to the action of heat in excess of air at temperature below its melting point. Roasting is usually carried out in a reverberatory furnace or in a blast furnace. During roasting:

(a) Volatile impurities like S, As, Sb etc. get oxidized and escape out as volatile gases SO_2, As_2O_3 and Sb_2O_3

(b) The sulphide ores decompose to their oxides evolving SO_2

(c) The moisture is removed. Mass becomes porous and thus it can easily be reduced. Roasting may be of many types.

Oxidising Roasting

In this type of roasting S, As, and Sb impurities are removed in the form of their volatile oxides as SO_2, As_2O_3 and Sb_2O_3 etc. due to combined action of heat and air. The ore is simultaneously converted into its oxides. This type of roasting is carried out for ores like copper pyrites, zinc blende and lead ores (PbS) etc.

$$2ZnS + 3O_2 \longrightarrow 2ZnO + 2SO_2$$

$$2PbS + 3O_2 \longrightarrow 2PbO + 2SO_2$$

CALCINATION

Another method of conversion of ore into metal oxide (oxidation) is called calcination. It is the process in which the ore is subjected to the action of heat at high temperature in the absence of air below its melting point. The process of calcination is carried out in the case of carbonate and hydrated ore. As a result of calcination :

(a) The moisture is removed.
(b) Gases may be expelled.
(c) Volatile impurities are removed.
(d) The mass becomes porous.
(e) Thermal decomposition of the ore takes place.

For example,

$$CaCO_3(\text{Limestone}) \longrightarrow Cao + CO_2 \uparrow$$

$$MgCO_3(\text{Magnesite}) \longrightarrow MgO + CO_2 \uparrow$$

$$MgCO_3.CaCO_3(\text{Dolomite}) \longrightarrow MgO + CaO + 2CO_2 \uparrow$$

$$CuCO_3.Cu(OH)_2(\text{Malachite}) \longrightarrow 2CuO + H_2O + CO_2 \uparrow$$

$$ZnCO_3(\text{Cala mine}) \longrightarrow ZnO + CO_2 \uparrow$$

$$2Fe_2O_3.3H_2O(\text{Limonite}) \longrightarrow 2Fe_2O_3 + 3H_2O \uparrow$$

The name calcination originated from the ore calcite which on thermal decomposition gives quick lime. Calcination is usually carried out in reverberatory furnace.

SMELTING – REDUCTION

Smelting is one of reduction method where the metal oxide is converted into metal. It is the process used for all operations where the metal is separated by fusion from the ore. The process of smelting is that in which ore is melted with a flux and often with a reducing agent, and it involves, calcination, roasting and reduction. In general, the process of separation of a metal or its sulphide mixture from its ore in fused state is called smelting. Smelting is generally carried out in a blast furnace and high temperature is produced by burning coal or by using electric energy. In smelting, the roasted or calcined ore is mixed with coke and then heated in a furnace.

As a result, (carbon and CO produced by the incomplete combustion) carbon reduces the oxide to the metal. For example, in the extraction of iron, haematite ore(Fe_2O_3) is smelted with coke and limestone (flux). As a result of reduction, iron is obtained in fused or molten state.

$$Fe_2O_3 + 3C \rightarrow 2Fe + 3CO$$

$$CaCO_3 \rightarrow CaO + CO_2$$

$$Fe_2O_3 + 3CO \rightarrow 2Fe + 3CO_2$$

$$CaO + SiO_2 \rightarrow CaSiO_3$$

$$\text{Flux} + \text{Gangue} \rightarrow \text{Slag}$$

Similarly, in the extraction of copper from copper pyrites, the ore is mixed with coke and heated in blast furnace (smelted). Infusible impurity FeO is converted to $FeSiO_3$(slag) and is removed. A mixture containing sulphide of copper and iron, called matte is formed in the molten state.

$$\underset{\text{Gangue}}{FeO} + \underset{\text{Flux}}{SiO_2} \rightarrow \underset{\text{Slag}}{FeSiO_3}$$

Other Example

$$ZnO + C \rightarrow Zn + CO$$

$$SnO_2 + 2C \rightarrow Sn + 2CO$$

$$MnO_2 + 2C \rightarrow Mn + 2CO$$

BESSEMERISATION

It is the process used for the manufacture of steel from pig iron. Steel is an alloy of carbon and iron and contains 0.15-1.5% of carbon with traces of sulphur, phosphorus, manganese and silicon as impurities. Depending upon the carbon content, steel are classified into three classes namely mild carbon steel (0.15-0.3%), medium carbon steel (0.3-0.8%) and high carbon steel (0.8-1.50).

The process was discovered by Henry Bessemer in England (1856). The principle involved in this process is that cold air blowed through refractory lined vessel known as converter containing molten pig iron at about 2 atmospheric pressure, oxidizing the impurities and simultaneously converting pig iron to steel.

This process mainly differs in the use of acidic and basic refractory linings of the converters. In this process low phosphorus pig iron (below 0.09%) is treated by acidic Bessemer process and high phosphorus pig iron (more than 1.5%) is treated in basic. The converter is a pear shaped furnace about 6m high and 3m in diameter. It is made of steel plates and is lined inside with silica or magnesia (MgO), depending upon the nature of impurities present in the pig iron.

If the impurities present in the pig iron are basic, e.g., manganese, a lining of silica brick is used and the process is known as acid Bessemer process. If impurities are acidic, e.g., sulphur, phosphorus etc., a basic lining of lime (CaO) or magnesia (MgO) is used in the converter and process is then known as basic Bessemer process.

The capacity of the converter is from 10-25 tonnes of charge at a time. The converter is mounted on shafts or trunnions, one of which is hollow and serves as a wind pipe and upon which the converter can rotate in any position. The converter is also provided with a number of holes at the bottom through which a hot blast of air can be introduced. The molten pig iron is mixed in mixers and then charged into converter. About 15-16 tonnes of iron can be charged at a time. The converter is first set in the horizontal position and after charging the converter is adjusted in vertical position. After charging a blast of cold air is admitted through the hole provided at the bottom at a pressure of about 2-3 kg/cm^3.

The blast is continued for about 15 minutes during which the impurities are oxidized. Mn is oxidized to MnO and Si is oxidized to SiO_2. Carbon is also oxidized to CO. The resulting oxides of Mn and Si (MnO and SiO_2) combine together to form slag of manganese silicate.

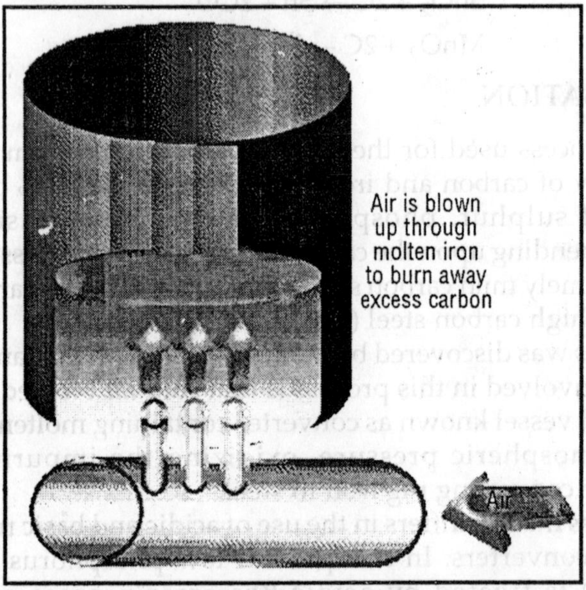

Fig. Bessemer Converter

PURIFICATION OF METALS

Electrolytic Refining

This is one of the most convenient and important method of refining and gives a metal of high purity. This method is applicable to many metals such as Cu, Ag, Pb, Au, Ni, Sn, Zn etc. The blocks of impure metal form the anode and thin sheets of pure metal form the cathode.

A solution of a salt of the metal is taken as an electrolyte. On passing an electric current through the solution pure metal dissolves from the anode and deposits on cathode. By this process, more metal ions undergo reduction and pure metal is deposited at the cathode.

The insoluble impurities either dissolve in the electrolyte or fall at the bottom and collect as anode mud. For example, in the refining of copper, impurities like Fe and Zn dissolve in the electrolyte, while Au, Ag and Pt are left behind as anode mud.

- *Copper:* During the electrolytic refining of a copper, a thick block of impure copper is made anode, and thin plate of pure copper is made cathode. Copper sulphate solution is used as an electrolyte. On passing electric current, following reactions take place:
- Cu^{2+} ions (from copper sulphate solution) go to the cathode (negative electrode), where they are reduced to copper, which gets deposited on the cathode.

$$Cu^{2+} + 2_e^- \rightarrow Cu$$

- Copper (of impure anode) forms copper ions, and these go into solution of electrolyte.

$$Cu \rightarrow Cu^{2+} 2_e^-$$

Thus, the net result is transfer of pure copper from anode to the cathode. Impurities like zinc, iron, etc., go into solution; while noble impurities like silver, gold, etc., are left behind as anode mud. Copper is refined to 99.98% pure copper by electrolytic refining.

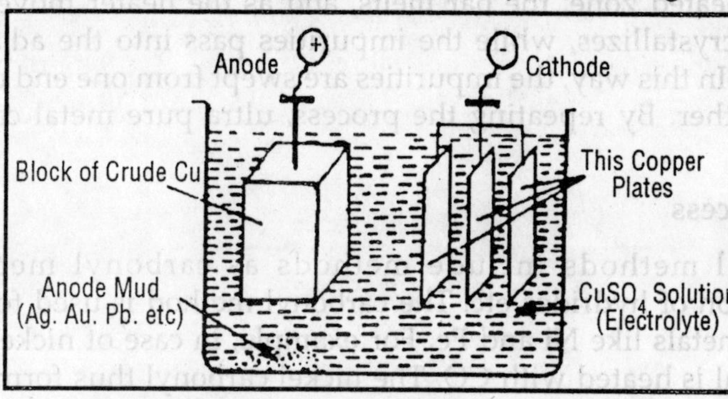

Fig. Electrolytic Refining of Copper

Zone Refining

This method is employed for preparing highly pure metal (such as silicon, tellurium, germanium), which are used as semiconductors. It is based on the principle that melting point of a substance is lowered by the presence of impurities.

Consequently, when an impure molten metal is cooled, crystals of the pure metal are solidified, and the impurities remain behind the remaining metal.

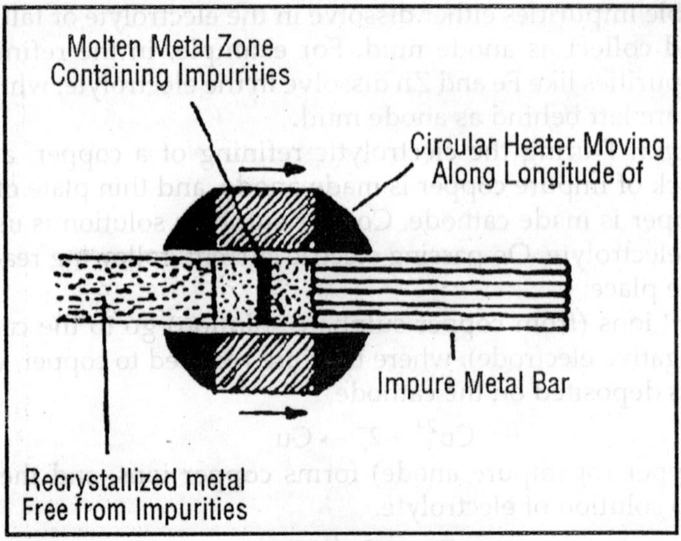

Fig. Zone Refining

The process consists in casting the impure metal in the form of a bar. A circular heater fitted around this bar is slowly moved longitudinally from one end to the other.

At the heated zone, the bar melts, and as the heater moves on, pure metal crystallizes, while the impurities pass into the adjacent molten part. In this way, the impurities are swept from one end of the bar to the other. By repeating the process, ultra pure metal can be obtained.

Mond's Process

Thermal methods include methods as carbonyl method, decomposition of hydrides etc. The carbonyl method is used for the refining of metals like Ni and Fe. For example, in case of nickel, the impure metal is heated with CO. The nickel carbonyl thus formed is then decomposed (after distilling off the impurities) to get pure nickel metal and CO. The process is known as

Mond's Process.

$$Ni + 4CO \rightarrow Ni + 4CO$$

It is based on the following facts:

- Only nickel (and not Cu, Fe, etc.) forms a volatile carbonyl, Ni(CO)4, when CO is passed over it at 50^0C.
- The nickel carbonyl decomposes at 180^0C to yield pure nickel.

METALLURGY OF ALUMINIUM

The extraction of aluminium from its ore, bauxite, including some economic and environmental issues. It finishes by looking at some uses of aluminium.

Extracting Aluminium from Bauxite

Aluminium is too high in the electrochemical series (reactivity series) to extract it from its ore using carbon reduction. The temperatures needed are too high to be economic. Instead, it is extracted by electrolysis. The ore is first converted into pure aluminium oxide by the Bayer Process, and this is then electrolysed in solution in molten cryolite - another aluminium compound. The aluminium oxide has too high a melting point to electrolyse on its own.

Aluminium Ore

The usual aluminium ore is *bauxite*. Bauxite is essentially an impure aluminium oxide. The major impurities include iron oxides, silicon dioxide and titanium dioxide.

Reaction with Sodium Hydroxide Solution

Crushed bauxite is treated with moderately concentrated sodium hydroxide solution. The concentration, temperature and pressure used depend on the source of the bauxite and exactly what form of aluminium oxide it contains. Temperatures are typically from 140°C to 240°C; pressures can be up to about 35 atmospheres. High pressures are necessary to keep the water in the sodium hydroxide solution liquid at temperatures above 100°C. The higher the temperature, the higher the pressure needed. With hot concentrated sodium hydroxide solution, aluminium oxide reacts to give a solution of sodium tetrahydroxoaluminate.

$$Al_2O_3 + 2NaOH + 3H_2O \rightarrow 2NaAl(OH)_4$$

The impurities in the bauxite remain as solids. For example, the other metal oxides present tend not to react with the sodium hydroxide solution and so remain unchanged. Some of the silicon dioxide reacts, but goes on to form a sodium aluminosilicate which precipitates out.

All of these solids are separated from the sodium tetrahydroxoaluminate solution by filtration. They form a "red mud" which is just stored in huge lagoons.

Precipitation of Aluminium Hydroxide

The sodium tetrahydroxoaluminate solution is cooled, and "seeded" with some previously produced aluminium hydroxide. This provides something for the new aluminium hydroxide to precipitate around.

$$NaAl(OH)_4 \rightarrow Al(OH)_3 + NaOH$$

Formation of Pure Aluminium Oxide

Aluminium oxide (sometimes known as alumina) is made by heating the aluminium hydroxide to a temperature of about 1100 - 1200°C.

$$2Al(OH)_3 \rightarrow Al_2O_3 + 3H_2O$$

Conversion of the Aluminium Oxide into Aluminium by Electrolysis

The aluminium oxide is electrolysed in solution in molten *cryolite*, Na_3AlF_6. Cryolite is another aluminium ore, but is rare and expensive, and most is now made chemically.

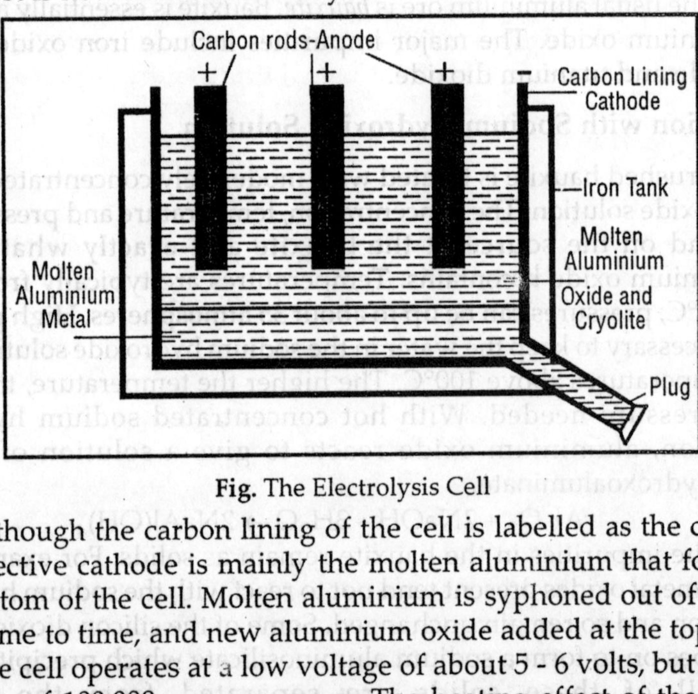

Fig. The Electrolysis Cell

Although the carbon lining of the cell is labelled as the cathode, the effective cathode is mainly the molten aluminium that forms on the bottom of the cell. Molten aluminium is syphoned out of the cell from time to time, and new aluminium oxide added at the top.

The cell operates at a low voltage of about 5 - 6 volts, but at huge currents of 100,000 amps or more. The heating effect of these large currents keeps the cell at a temperature of about 1000°C.

The Electrode Reactions

These are very complicated - in fact one source I've looked at says that they aren't fully understood. For chemistry purposes at this level, they are always simplified .This is the simplification:

Aluminium is released at the cathode. Aluminium ions are reduced by gaining 3 electrons.

$$Al^{3+} + 3e^- \rightarrow Al$$

Oxygen is produced initially at the anode.

$$2O^{2-} \rightarrow O_2 + 4e^-$$

However, at the temperature of the cell, the carbon anodes burn in this oxygen to give carbon dioxide and carbon monoxide. Continual replacement of the anodes is a major expense.

Economic Considerations

- The high cost of the process because of the huge amounts of electricity it uses. This is so high because to produce 1 mole of aluminium which only weighs 27 g we need 3 moles of electrons. We are having to add a lot of electrons (because of the high charge on the ion) to produce a small mass of aluminium (because of its low relative atomic mass).
- Energy and material costs in constantly replacing the anodes.
- Energy and material costs in producing the cryolite, some of which gets lost during the electrolysis.

Environmental Problems in Mining and Transporting the Bauxite

- Loss of landscape due to mining, processing and transporting the bauxite.
- Noise and air pollution (greenhouse effect, acid rain) involved in these operations.

Extracting Aluminium from the Bauxite

- Loss of landscape due to the size of the chemical plant needed, and in the production and transport of the electricity.
- Noise.
- Atmospheric pollution from the various stages of extraction. For example: carbon dioxide from the burning of the anodes (greenhouse effect); carbon monoxide (poisonous); fluorine (and fluorine compounds) lost from the cryolite during the electrolysis process (poisonous).
- Pollution caused by power generation (varying depending on how the electricity is generated.)
- Disposal of red mud into unsightly lagoons.
- Transport of the finished aluminium.

Recycling

- Saving of raw materials and particularly electrical energy by not having to extract the aluminium from the bauxite.

Recycling aluminium uses only about 5% of the energy used to extract it from bauxite.

- Avoiding the environmental problems in the extraction of aluminium from the bauxite.
- Not having to find space to dump the unwanted aluminium if it wasn't recycled.
- Offsetting these to a minor extent Energy and pollution costs in collecting and transporting the recycled aluminium.

USES OF ALUMINIUM

Aluminium is usually alloyed with other elements such as silicon, copper or magnesium. Pure aluminium isn't very strong, and alloying it adds to it strength.

- To make utensils
- To make automobile parts
- To make ice cube tray
- To make electrical wires
- To make roofs
- To use in reflecting telescope
- To obtain metals from Cr_2O_3, Mn_2O_3

Aluminium is especially useful because it

- Has a low density;
- Is strong when alloyed;
- Is a good conductor of electricity;
- Has a good appearance;
- Resists corrosion because of the strong thin layer of aluminium oxide on its surface. This layer can be strengthened further by anodising the aluminium.

Anodising essentially involves etching the aluminium with sodium hydroxide solution to remove the existing oxide layer, and then making the aluminium article the anode in an electrolysis of dilute sulphuric acid. The oxygen given of at the anode reacts with the aluminium surface, to build up a film of oxide up to about 0.02 mm thick.

As well as increasing the corrosion resistance of the aluminium, this film is porous at this stage and will also take up dyes. That means that we can make aluminium articles with the colour built into the surface.

PROPERTIES OF ALUMINIUM

- Silvery white in colour
- Density 2.6 g/cm^3
- Comparatively less hard

- Reducing property
- Conducts heat
- Reflects heat and light
- Conducts electricity

METALLURGY OF IRON

The Blast Furnace in the extraction of iron from iron ore, and the conversion of the raw iron from the furnace into various kinds of steel.

Extracting Iron from Iron ore using a Blast Furnace

The common ores of iron are both iron oxides, and these can be reduced to iron by heating them with carbon in the form of coke. Coke is produced by heating coal in the absence of air. Coke is cheap and provides both the reducing agent for the reaction and also the heat source.

Iron Ores

The most commonly used iron ores are *haematite* (US: hematite), Fe_2O_3, and *magnetite,* Fe_3O_4.

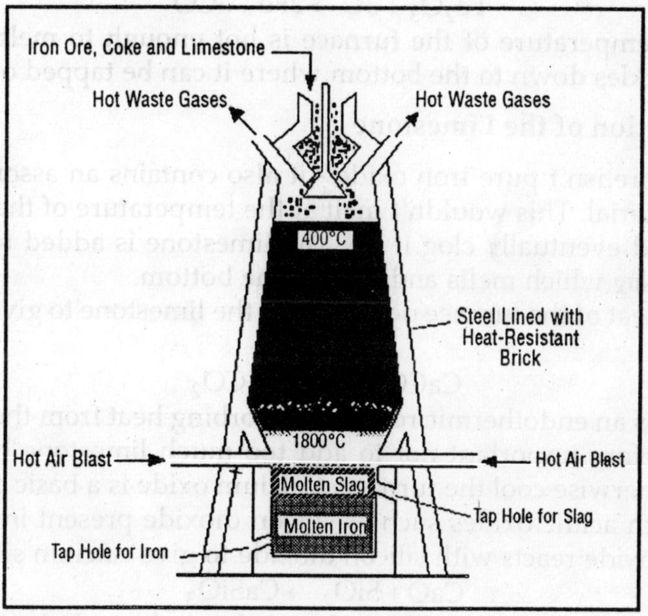

Fig. The Blast Furnace

The Heat Source

The air blown into the bottom of the furnace is heated using the hot waste gases from the top. Heat energy is valuable, and it is important not to waste any. The coke (essentially impure carbon) burns in the blast of hot air to form carbon dioxide - a strongly exothermic reaction. This reaction is the main source of heat in the furnace.

$$C + O_2 \rightarrow CO_2$$

The Reduction of the Ore

At the high temperature at the bottom of the furnace, carbon dioxide reacts with carbon to produce carbon monoxide.

$$C + CO_2 \rightarrow 2CO$$

It is the carbon monoxide which is the main reducing agent in the furnace.

$$Fe_2O_3 + 3CO \rightarrow 2Fe + 3CO_2$$

In the hotter parts of the furnace, the carbon itself also acts as a reducing agent. Notice that at these temperatures, the other product of the reaction is carbon monoxide, not carbon dioxide.

$$Fe_2O_3 + 3C \rightarrow 2Fe + 3CO$$

The temperature of the furnace is hot enough to melt the iron which trickles down to the bottom where it can be tapped off.

The Function of the Limestone

Iron ore isn't pure iron oxide - it also contains an assortment of rocky material. This wouldn't melt at the temperature of the furnace, and would eventually clog it up. The limestone is added to convert this into *slag* which melts and runs to the bottom.

The heat of the furnace decomposes the limestone to give calcium oxide.

$$CaCO_3 \rightarrow CaO + CO_2$$

This is an endothermic reaction, absorbing heat from the furnace. It is therefore important not to add too much limestone because it would otherwise cool the furnace. Calcium oxide is a basic oxide and reacts with acidic oxides such as silicon dioxide present in the rock. Calcium oxide reacts with silicon dioxide to give calcium silicate.

$$CaO + SiO_2 \rightarrow CaSiO_3$$

The calcium silicate melts and runs down through the furnace to form a layer on top of the molten iron. It can be tapped off from time to time as slag. Slag is used in road making and as "slag cement" - a final ground slag which can be used in cement, often mixed with Portland cement.

Cast Iron

The molten iron from the bottom of the furnace can be used as *cast iron*. Cast iron is very runny when it is molten and doesn't shrink much when it solidifies. It is therefore ideal for making castings - hence its name. However, it is very impure, containing about 4% of carbon. This carbon makes it very hard, but also very brittle. If we hit it hard, it tends to shatter rather than bend or dent. Cast iron is used for things like manhole covers, guttering and drainpipes, cylinder blocks in car engines, Aga-type cookers, and very expensive and very heavy cookware.

STEEL

Most of the molten iron from a Blast Furnace is used to make one of a number of types of steel. There isn't just one substance called steel - they are a family of alloys of iron with carbon or various metals.

Steel-making: The Basic Oxygen Process

Impurities in the iron from the Blast Furnace include carbon, sulphur, phosphorus and silicon. These have to be removed.

Heat Treatment of Steel

The properties of steel with the same composition can be varied by heating and cooling it under certain conditions. This is knows as heat treatment. Properties such as hardness, strength, ductility, malleability, and ability to resist corrosion are improved by heat treatment. Annealing, hardening and tempering are the important methods of heat treatment of steel. In annealing, steel is heated until it is bright red hot and allowed to cool slowly in air. Steel is made soft by this method. Hardening increases the hardness of steel.

In this method red hot steel is plunged into cold water or oil as quickly as possible. Hardened steel is heated again and is allowed to cool slowly in air. This method is called tempering.

Removal of Sulphur

Sulphur has to be removed first in a separate process. *Magnesium powder* is blown through the molten iron and the sulphur reacts with it to form magnesium sulphide. This forms a slag on top of the iron and can be removed.

$$Mg + s \rightarrow Mgs$$

Removal of Carbon

The still impure molten iron is mixed with scrap iron (from

recycling) and *oxygen* is blown on to the mixture. The oxygen reacts with the remaining impurities to form various oxides.The carbon forms carbon monoxide. Since this is a gas it removes itself from the iron! This carbon monoxide can be cleaned and used as a fuel gas. Elements like phosphorus and silicon react with the oxygen to form acidic oxides. These are removed using *quicklime (calcium oxide)* which is added to the furnace during the oxygen blow. They react to form compounds such as calcium silicate or calcium phosphate which form a slag on top of the iron.

TYPES OF IRON AND STEEL

Wrought Iron

If all the carbon is removed from the iron to give high purity iron, it is known as wrought iron. Wrought iron is quite soft and easily worked and has little structural strength. It was once used to make decorative gates and railings, but these days mild steel is normally used instead.

Mild Steel

Mild steel is iron containing up to about 0.25% of carbon. The presence of the carbon makes the steel stronger and harder than pure iron. The higher the percentage of carbon, the harder the steel becomes.

Mild steel is used for lots of things - nails, wire, car bodies, ship building, girders and bridges amongst others.

High Carbon Steel

High carbon steel contains up to about 1.5% of carbon. The presence of the extra carbon makes it very hard, but it also makes it more brittle. High carbon steel is used for cutting tools and masonry nails (nails designed to be driven into concrete blocks or brickwork without bending). We have to be careful with high carbon steel because it tends to fracture rather than bend if we mistreat it.

Special Steels

	iron mixed with	special properties	uses include
stainless steel	chromium and nickel	resists corrosion	cutlery, cooking utensils, kitchen sinks, industrial equipment for food and drink processing
titanium steel	titanium	withstands high temperatures	gas turbines, spacecraft
manganese steel	manganese	very hard	rock-breaking machinery, some railway track (e.g. points), military helmets

ENVIRONMENTAL CONSIDERATIONS

Environmental Problems in Mining and Transporting the Raw Materials

- Loss of landscape due to mining, processing and transporting the iron ore, coke and limestone.
- Noise and air pollution (greenhouse effect, acid rain) involved in these operations.

Extracting Iron from the Ore

- Loss of landscape due to the size of the chemical plant needed.
- Noise.
- Atmospheric pollution from the various stages of extraction. For example: carbon dioxide (greenhouse effect); carbon monoxide (poisonous); sulphur dioxide from the sulphur content of the ores (poisonous, acid rain).
- Disposal of slag, some of which is just dumped.
- Transport of the finished iron.

Recycling

- Saving of raw materials and energy by not having to first extract the iron from the ore.
- Avoiding the pollution problems in the extraction of iron from the ore.
- Not having to find space to dump the unwanted iron if it wasn't recycled.
- (Offsetting these to a minor extent) Energy and pollution costs in collecting and transporting the recycled iron to the steel works.

METALLURGY OF COPPER

The most common source of copper ore is the mineral chalcopyrite ($CuFeS_2$), which accounts for about 50% of copper production. The process of copper extraction from chalcopyrite ore into pure metal.

For economic and environmental reasons, many of the byproducts of extraction are reclaimed. Sulfur dioxide gas, for example, is captured and turned into sulfuric acid – which is then used in the extraction process. The extraction of copper from its ores, its purification by electrolysis, and some of its uses.

Extracting Copper from its Ores

The method used to extract copper from its ores depends on the nature of the ore. Sulphide ores such as chalcopyrite are converted to copper by a different method from silicate, carbonate or sulphate ores.

Getting Copper from Chalcopyrite, $CuFeS_2$

Chalcopyrite (also known as copper pyrites) and similar sulphide ores are the commonest ores of copper. The ores typically contain low percentages of copper and have to be concentrated by, for example, froth flotation before refining.

The Process

The concentrated ore is heated strongly with silicon dioxide (silica) and air or oxygen in a furnace or series of furnaces.

- The copper(II) ions in the chalcopyrite are reduced to copper(I) sulphide.
- The iron in the chalcopyrite ends up converted into an iron(II) silicate slag which is removed.
- Most of the sulphur in the chalcopyrite turns into sulphur dioxide gas. This is used to make sulphuric acid via the Contact Process.

An overall equation for this series of steps is:

$$2CuFeS_2 + 2SiO_2 + 4O_2 \longrightarrow Cu_2S + 2FeSiO_3 + 3SO_2$$

The copper(I) sulphide produced is converted to copper with a final blast of air.

$$Cu_2S + O_2 \longrightarrow 2Cu + SO_2$$

The end product of this is called *blister copper* - a porous brittle form of copper, about 98 - 99.5% pure.

Exploring the redox processes in this reaction

$$Cu_2S + O_2 \longrightarrow 2Cu + SO_2$$

- In the copper(I) sulphide, the copper is +1 and the sulphur -2.
- The oxidation states of the elements oxygen (in the gas) and copper (in the metal) are 0.
- In sulphur dioxide, the oxygen has an oxidation state of -2 and the sulphur +4.

That means that both the copper and the oxygen have been reduced (decrease in oxidation state). The sulphur has been oxidised (increase in oxidation state). The reducing agent is therefore the sulphide ion in the copper(I) sulphide. The other reaction is more difficult to deal with, because we can't work out all of the oxidation states by following the simple rules - there are too many variables in some of the substances. We have to use some chemical knowledge as well.

$$2CuFeS_2 + 2SiO_2 + 4O_2 \longrightarrow Cu_2S + 2FeSiO_3 + 3SO_2$$

In the $CuFeS_2$, we would have to know that the copper and iron are both in oxidation state +2, for example. We would also have to know that the oxidation state of the silicon remains unchanged at +4.

So use that information to work out what has been oxidised and what reduced in this case!

We should find that copper has been reduced from +2 to +1; oxygen (in the gas) has been reduced from 0 to -2 (oxygen in the SiO_2 is unchanged); and three of the four sulphurs on the left-hand side have been oxidised from -2 to +4 (the other is unchanged). Once again, the sulphide ions are acting as the reducing agent.

EXTRACTING COPPER FROM OTHER ORES

Copper can be extracted from non-sulphide ores by a different process involving three separate stages:

- Reaction of the ore (over quite a long time and on a huge scale) with a dilute acid such as dilute sulphuric acid to produce a very dilute copper(II) sulphate solution.
- Concentration of the copper(II) sulphate solution by solvent extraction.

The very dilute solution is brought into contact with a relatively small amount of an organic solvent containing something which will bind with copper(II) ions so that they are removed from the dilute solution. The solvent mustn't mix with the water.

The copper(II) ions are removed again from the organic solvent by reaction with fresh sulphuric acid, producing a much more concentrated copper(II) sulphate solution than before.

- Electrolysis of the new solution. Copper(II) ions are deposited as copper on the cathode (for the electrode equation.

The anodes for this process were traditionally lead-based alloys, but newer methods use titanium or stainless steel.

The cathode is either a strip of very pure copper which the new copper plates on to, or stainless steel which it has to be removed from later.

PURIFICATION OF COPPER

When copper is made from sulphide ores by the first method it is impure. The blister copper is first treated to remove any remaining sulphur (trapped as bubbles of sulphur dioxide in the copper - hence "blister copper") and then cast into anodes for refining using electrolysis.

Electrolytic Refining

The purification uses an electrolyte of copper(II) sulphate solution, impure copper anodes, and strips of high purity copper for the cathodes.

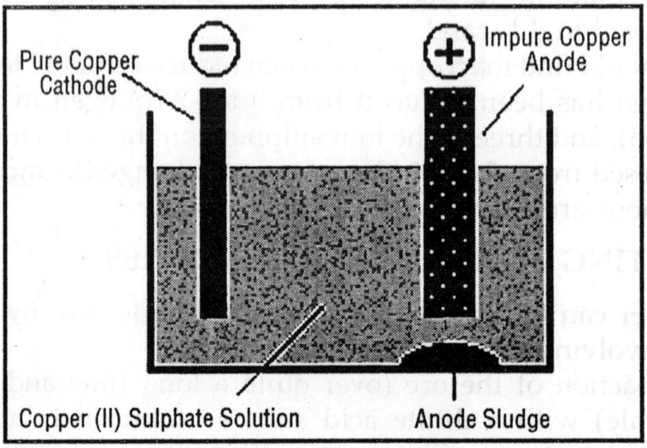

Fig. Electrolytic Refining of Copper

At the cathode, copper(II) ions are deposited as copper.

$$Cu^{2+}(aq) + 2e^{-} \longrightarrow Cu_{(s)}$$

At the anode, copper goes into solution as copper(II) ions.

$$Cu_{(s)} \longrightarrow Cu^{2+}(aq) + 2e^{-}$$

For every copper ion that is deposited at the cathode, in principle another one goes into solution at the anode. The concentration of the solution should stay the same.

All that happens is that there is a transfer of copper from the anode to the cathode. The cathode gets bigger as more and more pure copper is deposited; the anode gradually disappears.

Any metal in the impure anode which is below copper in the electrochemical series (reactivity series) doesn't go into solution as ions. It stays as a metal and falls to the bottom of the cell as an "anode sludge" together with any unreactive material left over from the ore. The anode sludge will contain valuable metals such as silver and gold.

Metals above copper in the electrochemical series (like zinc) *will* form ions at the anode and go into solution. However, they won't get discharged at the cathode provided their concentration doesn't get too high.

The concentration of ions like zinc will increase with time, and the concentration of the copper(II) ions in the solution will fall. For every zinc ion going into solution there will obviously be one fewer copper ion formed. The copper(II) sulphate solution has to be continuously purified to make up for this.

USES OF COPPER

Amongst other things copper is used for:

- Electrical wiring. It is a very good conductor of electricity and is easily drawn out into wires.
- Domestic plumbing. It doesn't react with water, and is easily bent into shape.
- Boilers and heat exchangers. It is a good conductor of heat and doesn't react with water.
- Making brass. Brass is a copper-zinc alloy. Alloying produces a metal harder than either copper or zinc individually. Bronze is another copper alloy - this time with tin.

ALLOY

An alloy is a partial or complete solid solution of one or more elements in a metallic matrix. Complete solid solution alloys give single solid phase microstructure, while partial solutions give two or more phases that may be homogeneous in distribution depending on thermal (heat treatment) history. Alloys usually have different properties from those of the component elements.

Alloying one metal with other metal(s) or non metal(s) often enhances its properties. For instance, steel is stronger than iron, its primary element. The physical properties, such as density, reactivity, Young's modulus, and electrical and thermal conductivity, of an alloy may not differ greatly from those of its elements, but engineering properties, such as tensile strength and shear strength may be substantially different from those of the constituent materials.

This is sometimes due to the sizes of the atoms in the alloy, since larger atoms exert a compressive force on neighboring atoms, and smaller atoms exert a tensile force on their neighbors, helping the alloy resist deformation.

Alloys may exhibit marked differences in behaviour even when small amounts of one element occur. For example, impurities in semi-conducting ferromagnetic alloys lead to different properties, as first predicted by White, Hogan, Suhl, Tian Abrie and Nakamura. Some alloys are made by melting and mixing two or more metals. Brass is an alloy made from copper and zinc. Bronze, used for bearings, statues, ornaments and church bells, is an alloy of tin and copper. Unlike pure metals, most alloys do not have a single melting point. Instead, they have a melting range in which the material is a mixture of solid and liquid phases. The temperature at which melting begins is called the solidus and the temperature when melting is complete is called the liquidus. However, for most alloys there is a particular proportion of constituents which give them a single melting point or (rarely) two. This is called the alloy's eutectic mixture.

CLASSIFICATION OF ALLOY

Alloys can be classified by the number of their constituents. An alloy with two components is called a binary alloy; one with three is a ternary alloy, and so forth. Substitution alloys or Interstitial alloys, depending on their method of formation.

In substitution alloys, the atoms of the components are approximately the same size and the various atoms are simply substituted for one another in the crystal structure. An example of a (binary) substitution alloy is brass, made up of copper and zinc. Interstitial alloys occur when the atoms of one component are substantially smaller than the other and the smaller atoms fit into the spaces (interstices) between the larger atoms.

PROPERTIES AND USES OF ALLOYS

BRASS

Brass is any alloy of copper and zinc; the proportions of zinc and copper can be varied to create a range of brasses with varying properties. In comparison, bronze is principally an alloy of copper and tin. Despite this distinction, some types of brasses are called bronzes. Brass is a substitutional alloy. It is used for decoration for its bright gold-like appearance; for applications where low friction is required such as locks, gears, bearings, ammunition, and valves; for plumbing and electrical applications; and extensively in musical instruments such as horns and bells for its acoustic properties.

Brass has a muted yellow colour, somewhat similar to gold. It is relatively resistant to tarnishing, and is often used as decoration and for coins. In antiquity, polished brass was often used as a mirror.Brass has likely been known to humans since prehistoric times, even before zinc itself was discovered.

It was produced by melting copper together with calamine, a zinc ore. In the German village of Breinigerberg, an ancient Roman settlement was discovered where a calamine ore mine existed. During the melting process, the zinc is extracted from the calamine and mixes with the copper.

Pure zinc, on the other hand, has too low a boiling point to have been produced by ancient metalworking techniques. The many references to 'brass' appearing throughout the King James Bible are thought to signify another bronze alloy, or copper, rather than the strict modern definition of 'brass'.

Properties of Brass

The malleability and acoustic properties of brass have made it the metal of choice for brass musical instruments such as the trombone, tuba, trumpet, euphonium, and the French horn. Even though the saxophone is classified as a woodwind instrument and the harmonica is a free reed aerophone, both are also often made from brass. In organ pipes designed as "reed" pipes, brass strips are used as the "reeds".

Brass has higher malleability than copper or zinc. The relatively low melting point of brass (900 to 940°C, depending on composition) and its flow characteristics make it a relatively easy material to cast. By varying the proportions of copper and zinc, the properties of the brass can be changed, allowing hard and soft brasses. The density of brass is approximately 8.4 g/cm^3.

Today almost 90% of all brass alloys are recycled. Because brass is nonmagnetic, it can be separated from ferrous scrap by passing the scrap near a powerful magnet. Brass scrap is collected and transported to the foundry where it is melted and recast into billets. Billets are heated and extruded into the desired form and size.

Aluminum makes brass stronger and more corrosion resistant. Aluminum also causes a highly beneficial hard layer of aluminium oxide (Al_2O_3) to be formed on the surface that is thin, transparent and self healing. Tin has a similar effect and finds its use especially in sea water applications (naval brasses). Combinations of iron, aluminum, silicon and manganese make brass wear and tear resistant. A well known alloy used in the automotive industry is 'LDM C673', where the combination of manganese and silicon leads to a strong and resistant brass.

Applications of Brass

The so called dezincification resistant (DZR) brasses, like alloy 'LDM G563', are used where there is a large corrosion risk and where normal brasses do not meet the standards. Applications with high water temperatures, chlorides present or deviating water qualities (soft water) play a role.

DZR-brass is excellent in water boiler systems. This brass alloy must be produced with great care, with special attention placed on a balanced composition and proper production temperatures and parameters to avoid long-term failures.

The copper in brass makes brass germicidal, via the oligodynamic effect. For example, brass doorknobs disinfect themselves of many bacteria within eight hours. This effect is important in hospitals, but useful in many contexts.

Brass door hardware is generally lacquered when new, which prevents tarnishing of the metal for a few years when located outside (and indefinitely when located indoors). After this most manufacturers recommend that the lacquer is removed (e.g. with paint stripper) and the items regularly polished to maintain a bright finish.

Unlacquered brass weathers more attractively than brass with deteriorated lacquer, even if polishing is not carried out. Freshly polished brass is similar to gold in appearance, but becomes more reddish within days of exposure to the elements. A traditional polish is Brasso.

Brass was used to make fan blades, fan cages and motor bearings in many antique fans that date before the 1930s. Brass can be used for fixings for use in cryogenic systems, however its use is not limited to this.

Due to its natural heat conducting properties and its availability, brass was used to create the infamous Brazen Bull, a torture and execution device in ancient Greece.

Types of Brass

- Admiralty brass contains 30% zinc and 1% tin which inhibits dezincification in most environments.
- Alpha brasses (Prince's metal), with less than 35% zinc, are malleable, can be worked cold, and are used in pressing, forging, or similar applications. They contain only one phase, with face-centered cubic crystal structure.
- Alpha-beta brass (Muntz metal), also called duplex brass, is 35-45% zinc and is suited for hot working. It contains both á and â′ phase; the â′-phase is body-centered cubic and is harder and stronger than á. Alpha-beta brasses are usually worked hot.
- Aluminium brass contains aluminium, which improves its corrosion resistance. Used in Euro coins (Nordic gold).
- Arsenical brass contains an addition of arsenic and frequently aluminium and is used for boiler fireboxes.

- Beta brasses, with 45-50% zinc content, can only be worked hot, and are harder, stronger, and suitable for casting.
- Cartridge brass is a 30% zinc brass with good cold working properties.
- Common brass, or rivet brass, is a 37% zinc brass, cheap and standard for cold working.
- DZR brass is Dezincification resistant Brass with a small percentage of Arsenic.
- Gilding metal is the softest type of brass commonly available. An alloy of 95% copper and 5% zinc, gilding metal is typically used for ammunition components.
- High brass, contains 65% copper and 35% zinc, has a high tensile strength and is used for springs, screws, rivets.
- Leaded brass is an alpha-beta brass with an addition of lead. It has excellent machinability.
- Low brass is a copper-zinc alloy containing 20% zinc with a light golden colour, excellent ductility and is used for flexible metal hoses and metal bellows.
- Naval brass, similar to admiralty brass, is a 40% zinc brass and 1% tin.
- Red brass, while not technically brass, is an American term for CuZnSn alloy known as gunmetal.
- Rich low brass contains 85% copper 15% zinc often used in jewelry applications .
- White brass contains more than 50% zinc and is too brittle for general use.
- Yellow brass is an American term for 33% zinc brass.

BRONZE

Bronze is any of a broad range of copper alloys, usually with tin as the main additive, but sometimes with other elements such as phosphorus, manganese, aluminum, or silicon. It was particularly significant in antiquity, giving its name to the Bronze Age. "Bronze," in turn, is perhaps ultimately taken from the Persian word "berenj," meaning "brass".

Bronze was significant to any culture that encountered it. It was one of the most innovative alloys of mankind. Tools, weapons, armour, and various building materials like decorative tiles made of bronze were harder and more durable than their stone and copper ("Chalcolithic") predecessors. In early use, the impurity arsenic sometimes created a superior alloy; this is termed arsenical bronze.

The two ores are rarely found together (exceptions include one ancient site in Thailand and one in Iran), so serious bronze work has always involved trade. In Europe, the major source for tin was Great Britain's deposits of ore in Cornwall. Phoenician traders visited Great Britain to trade goods from the Mediterranean for tin.

Though bronze is stronger (harder) than wrought iron, the Bronze Age gave way to the Iron Age; this happened because iron was easier to find. Bronze was still used during the Iron Age, but for many purposes the weaker wrought iron was found to be sufficiently strong. Archaeologists suspect that a serious disruption of the tin trade precipitated the transition.

The population migrations around 1200 – 1100 BC reduced the shipping of tin around the Mediterranean, limiting supplies and raising prices. As ironworking improved, iron became cheaper, and cultures learned how to make steel, which is stronger than bronze and holds a sharper edge longer.

Properties of Bronze

With the exception of steel, bronze is superior to iron in nearly every application. It is considerably less brittle than iron. Bronze only oxidizes superficially; once the surface oxidizes, the thin oxide layer protects the underlying metal from further corrosion. Copper-based alloys have lower melting points than steel or iron, and are more readily produced from their constituent metals.

They are generally about 10 per cent heavier than steel, although alloys using aluminum or silicon may be slightly less dense. Bronzes are softer and weaker than steel, e.g. bronze springs are less stiff (and so store less energy) for the same bulk. Bronze resists corrosion (especially seawater corrosion) and metal fatigue better than steel and also conducts heat and electricity better than most steels. The cost of copper-base alloys is generally higher than that of steels but lower than that of nickel-base alloys such as stainless steel.

Copper and its alloys have a huge variety of uses that reflect their versatile physical, mechanical, and chemical properties. Some common examples are the high electrical conductivity of pure copper, the excellent deep-drawing qualities of cartridge case brass, the low-friction properties of bearing bronze, the resonant qualities of bell bronze, and the resistance to corrosion by sea water of several bronze alloys.

In the twentieth century, silicon was introduced as the primary alloying element, creating an alloy with wide application in industry and the major form used in contemporary statuary. Aluminum is also used for the structural metal aluminum bronze.

It is also widely used for cast metal sculpture. Many common bronze alloys have the unusual and very desirable property of expanding slightly just before they set, thus filling in the finest details of a mould. Bronze parts are tough and typically used for bearings, clips, electrical connectors and springs.

Bronze also has very little metal-on-metal friction, which made it invaluable for the building of cannon where iron cannonballs would otherwise stick in the barrel. It is still widely used today for springs, bearings, bushings, automobile transmission pilot bearings, and similar fittings, and is particularly common in the bearings of small electric motors. Phosphor bronze is particularly suited to precision-grade bearings and springs. It is also used in guitar and piano strings.

Bronze is typically 88% copper and 12% tin. Alpha bronze consists of the alpha solid solution of tin in copper. Alpha bronze alloys of 4–5% tin are used to make coins, springs, turbines and blades.

Commercial bronze (otherwise known as brass) is 90% copper and 10% zinc, and contains no tin. It is stronger than copper and it has equivalent ductility. It is used for screws and wires.

Unlike steel, bronze struck against a hard surface will not generate sparks, so it (along with beryllium copper) is used to make hammers, mallets, wrenches and other durable tools to be used in explosive atmospheres or in the presence of flammable vapours.

MONEL METAL

Monel is a trademark of Special Metals Corporation for a series of nickel alloys, primarily composed of nickel (up to 67%) and copper, with some iron and other trace elements. Monel was created by David H. Browne, chief metallurgist for International Nickel Co. Monel alloy 400 is binary alloy of the same proportions of nickel and copper as is found naturally in the nickel ore from the Sudbury (Ontario) mines.

Properties of Monel Metal

Compared to steel, Monel is very difficult to machine as it work-hardens instantly with heat and does not harden into a constant matrix. It needs to be turned and worked at slow speeds and low feed rates. It is resistant to corrosion and acids, and some alloys can withstand a fire in pure oxygen.

It is commonly used in applications with highly corrosive conditions. Small additions of aluminium and titanium form an alloy (K-500) with the same corrosion resistance but with much greater strength due to gamma prime formation on aging. Monel is typically much more expensive than stainless steel.

- Monel has a specific gravity of 8.82.

- Monel has an electrical conductivity of approximately 3.6% IACS.

USES OF MONEL METAL

Marine Applications

Monel's corrosion resistance makes it ideal for marine applications such as piping systems, pump shafts, trolling wire, and strainer baskets. Some alloys are completely non-magnetic and are used for anchor cable aboard minesweepers, housings for magnetic-field measurement equipment, and have applications in the oil drilling industry.

In recreational boating, Monel wire is used to seize shackles for anchor rodes, Monel is used for water and fuel tanks, and for under water applications. It is also used for propeller shafts and for keel bolts.

Musical Instruments

Monel is used as the material for valve pistons in some higher quality musical instruments such as trumpets, tubas and french horn rotors. RotoSound introduced the use of Monel for electric bass strings in 1962, and these strings have been used by numerous artists, including. The Who, Sting, John Deacon, and John Paul Jones. Monel was in use in the early 1930s by other musical string manufacturers, such as Gibson Guitar Corporation, who continue to offer them for mandolin as the Sam Bush signature set.

Other

Monel is often used for kitchen sinks and in the frames of eyeglasses. It has also been used for firebox stays in fire-tube boilers.Parts of the Clock of the Long Now, which is intended to run for 10,000 years, are made from Monel because of the corrosion resistance without the use of precious metals. Monel was used for much of the exposed metal used in the interior of the Bryn Athyn Cathedral. This included large decorative screens, doorknobs, etc. Monel also has been used as roofing material in buildings such as Penn Station.

MAGNALIUM

Magnalium refers to an alloy of aluminium with 5-50% magnesium. Trace amounts of other elements are added to improve hardness. It finds use in engineering and pyrotechnics.

Properties of Magnalium

Alloys with smaller amounts of magnesium (about 5%) exhibit greater strength, greater corrosion resistance, and lower density than pure aluminium. Such alloys are also more workable and easier to weld than pure aluminium. Alloys with high amounts of magnesium (around 50%) are brittle and more susceptible to corrosion than aluminium.

Uses of Magnalium

Although they are generally more expensive than aluminium, the high strength, low density, and greater workability of alloys with low amounts of magnesium leads to their use in aircraft and automobile parts. Alloys with about 50% magnesium are brittle and corrode easily, which makes them unsuitable for most engineering uses. However, these alloys are flammable when powdered, are more resistant to corrosion than pure magnesium, and are more reactive than pure aluminium and are therefore used in pyrotechnics as a metal fuel and to produce sparks.

DURALUMIN

Duralumin is the trade name of one of the earliest types of age-hardenable aluminium alloys. The main alloying constituents are copper, manganese and magnesium. A commonly used modern equivalent of this alloy type is AA2024, which contains (in wt.%) 4.4% copper, 1.5% magnesium and 0.6% manganese. Typical yield strength is 450 MPa, with variations depending on the composition and temper.

Duralumin was developed by the German metallurgist Alfred Wilm at Dürener Metallwerke Aktien Gesellschaft. In 1903, Wilm discovered that after quenching, an aluminium alloy containing 4% copper would slowly harden when left at room temperature for several days. Further improvements led to the introduction of Duralumin in 1909. The name is obsolete today, and mainly used in popular science to describe the Al-Cu alloy system, or 2000 series as designated by the Aluminum Association. Its first use was rigid airship frames. Its composition and heat treatment were a wartime secret. With this new rip-resistant mixture, duralumin quickly spread throughout the aircraft industry in the early 1930s, where it was well suited to the new monocoque construction techniques that were being introduced at the same time. Duralumin also is popular for use in precision tools such as levels because of its light weight and strength. Although the addition of copper improves strength, it also makes these alloys susceptible to corrosion. For sheet products, corrosion resistance can be greatly enhanced by metallurgical bonding of a high-purity aluminium surface layer. These sheets are referred to as alclad, and are commonly used by the aircraft industry.

Chapter 6

Plant Genetic Resources

Plant genetic resources are the most valuable and essential basic raw materials to meet the current and future needs of crop improvement programmes. A wider genetic base, thus, assumes priority in plant breeding research aimed at developing new varieties for increased crop production. This diversity comprises native landraces, local selections, elite cultivars and wild relatives of crop plants.

The collection and conservation of this diversity in a systematic manner is the primary responsibility of all plant genetic resources institutes/centres: In fact, in the national context, in all those countries which hold native genetic diversity, the scientists, planners and policy makers have a challenge as to how best the existing genetic diversity can be preserved and utilised for the benefit of mankind. Plant genetic resources are thus our heritage which need conservation for posterity.

Man's interest in agriculture started about 10,000 years ago and, during this long period, transition from 'gathering' to 'growing' of plants occurred. In this process, a wide array of crop variability got generated by natural means and through both conscious or unconscious selection.

Gradually, a new wealth of variability also got generated/adapted and diversified by crop introductions in the exotic environment or through migration of human population. Associated with this process was the keenness of human mind to explore the rich global diversity of plant wealth in general so as to judiciously tap the potential of useful flora.

Historically, mankind has used only about 5,000 plant species worldwide to meet food and other needs. This number is just a fraction of the total world flora. With population growth, we are increasingly dependent on most productive plants. Today, only about 150 plant species are important in meeting the food (calories) needs of humans worldwide.

The short list of the most commonly used foods is what stands between us and the estimated worldwide carrying capacity in pre-agricultural times :

Cereals	: Wheat, rice, maize, barley, oats, sorghum, millets, rye.
Oilseeds	: Coconut, cotton seed, sunflower.
Legumes	: Soybean, peanut, common beans (*Phaseolus* spp.), pea, chickpea, cowpea.
Root crops	: Potato, sweet potato, cassava, yam and taro.
Sugar crops	: Sugarcane, sugarbeet.
Vegetables	: Tomato, cabbage, onion, squash.
Fruits	: Banana, orange, apple, pear, melon, mango.

The cereals represent twice the tonnage of all the other crops (if the potato with its high water content is discounted) and nearly three times the calories. Both tea and coffee are major world crops but these caffeine containing beverages are not included because they add essentially no calories to human nutrition.

This list is primarily a calorie list and does not recognise the important role of low calorie vegetables and fruits in supplying vitamins, minerals and proteins to human nutrition. The list also does not include regional foods that locally may supply more than half of the calories consumed. In addition, pasture forages and fibre crops have been omitted .

Hence, greater dependence on fewer plant species, 20 to 30 in global context gradually, has resulted in the loss of native genetic resources which are otherwise essential as building blocks of genetic diversity. In order to safeguard and conserve this diversity, the large share of which is held by the developing world, different countries/ national and international organisations have in recent past shown their concern for proper conservation of genetic resources.

The cultivated plants truly form the very basis of modern civilization and their domestication in historical context constitutes the agricultural revolution, the signs of which are still visible in traditional cultivation sites in the developing world. Emphasis is laid on some of these findings/concepts relating to the origin of agriculture, the process of plant domestication and the geographical regions where genetic richness in plant resources occurs. Further, the importance to safeguard and conserve such- genepools has been highlighted.

CHANGE FROM NOMADIC LIFE TO SETTLED AGRICULTURE

To quote the great scientific philosopher, 'the genesis of cultivated plants is a true fairy tale of genetics as if coming of civilization had been blessed in advance by the spirit of the abbot Gregor Mendel'. The change from nomadic life to settled agriculture has had a fascinating past.

These are some of the basic questions, answers to which can apprise the students of botany and agriculture of the value of plant genetic resources our 'Mother Earth' is so greatly endowed with. These points of plant genetic diversity, largely in the Indian perspective.

Conclusions drawn from circumstantial evidences show that agriculture started about 10,000 years ago. A well known hypothesis for the origin of agriculture is the Rubbish-Heap hypothesis. It says that early humans gathered nutritious roots and seeds for their food. Such plants actively colonized the bare areas around their dwellings, which were rich with the discarded rubbish. This natural process was obviously based on man's known food gathering activities and selection of only those useful plants which he found most suitable in tune with local habitats.

The plants selected by man must have been preadapted to agriculture and they must be of particular interest to man because of their large food reserves. Three stages seemed to have operated in this continuous process. In the first stage, the preadapted wild plants with weedy tendencies and large reserves of food began to colonize the open ground or kitchen middens around man's dwellings.

Also, people probably accidentally dropped seeds from the natural habitats on the same ground and thus reinforced the process. In the second stage, perhaps the seeds were regularly harvested from the plots around the dwellings, and fenced to protect them from domesticated cattle and wild herbivores.

At this stage, the early man was also selecting mutants for increased yields, palatability and other desirable traits, and the plants might have evolved a series of better adapted types that were able to take advantage of the richer soil conditions. The third stage of sowing/ placement of seeds in soil at the right time and carefully guarding them at all stages until harvest must have come very late at a time when man really began to know his plants.

This stage marks the end of pre-agricultural phase and perhaps led to considerable increase of human population as a result of increased food supplies. From here, the human families would have begun to move away from their original areas, eventually reaching regions where the wild progenitors of crop plants no longer grew.

These families could continue to harvest the food plants in the new region only if they had taken some seeds with them. As the population continued to spread, the new tradition of keeping and saving seeds, a tradition born out of necessity, may have been established. This explanation can obviously account for both seed and tuber/root crops.

DYNAMICS OF PLANT DOMESTICATION

Domestication is an evolutionary process operating under the influence of human activities. Being evolutionary, obviously it is relatively a slow process and exhibits gradual progression from the wild state to a state of incipient domestication.

Diverse forms that differ more and more from their progenitors develop. The possible changes in plant species due to domestication are listed below:

- Adapting to a greater diversity of environments and a wider geographical range.
- Different/specific ecological preference.
- Flowering and fruiting simultaneously/uniformly.
- Lack of shattering or scattering of seeds and sometimes may have lost the dispersal mechanism completely.
- Increased size of fruits and seeds which often reduces the dispersal efficiency.
- Change from a perennial to annual habit.
- Loss of seed dormancy.
- Loss of photoperiodic response.
- Lack of normal pollinating organs.
- Change in breeding system may result from a change in flower morphology, or a change from self-incompatibility to self-compatibility.
- Loss of defensive adaptations, such as hairs, spines, thorns, etc.
- Loss of protective coverings and sturdiness.
- Improvement towards palatability, chemical composition rendering them more likely to be eaten by animals.
- Increased susceptibility for diseases and pests.
- Developing of seedless parthenocarpic fruits.
- Multiplied vegetatively.

Cultivation practices adopted by man also had a significant role in the domestication process, the cultivated field presenting a different environment from the wild habitat. The crop evolutionary process obviously includes changes as affected by changed environment of a cultivated field.

Not only this, the selection pressure associated with cultivation practices also results in production of weedy races plants which are competitive with cultivated races but retain some important characters of the wild races and consequent occasional crossing between the two, leading to a setting up of differentiation-hybridization cycle and release of more potential variability.

This is one remarkably elegant evoloutionary process wherein barriers to geneflow maintain identity of the two types and, at the same time, limited exchange of genes releases variability. Deliberate selection practices by man from the released variability provide a new order of selection pressure making the population an array of deliberately chosen components.

The total potential range of variation is also fragmented into landrace populations or primitive cultivars. Different cultivars are grown for different purposes or to fit different ecological niches of the agricultural system. The man selects glutinous and non-glutinous rice, long and short grained rice, aromatic rice, etc. In maize, the man's selection is for popping, boiling, eating tender cob, flour quality, etc.

Broadly, three factors that operate in the selection process of domesticated species are:

- Selection of desirable traits by the cultivator while sowing;
- Changed micro-environment through cultivation practices;
- Differentiation - hybridization cycles between crop-weed pairs and man's selection from them.

So the dynamics of domestication has resulted in great morphological changes without substantial change in the genetic background. However, speciation rarely occurs under domestication. Under domestication, modification(s) induced ultimately lead to the end products which are generally radically different in appearance from their wild progenitors.

Thus, most domesticated plants and all the food plants are, by and large, the product of a long selection process. As already elaborated, we call this evolutionary process as domestication. In the process of domestication, food plants have quite literally crossed a threshold.

Their survival is keyed to preparation of the ground, decreased competition with other plants, sowing of the seed in the proper season, protection of plants during their growth and finally collecting seed. Thus, the process of domestication has 'tamed' these plants to make them dependent on humans. On the global scene, the human population has enormously increased such that we are held captive by our domesticated food plants, that is we are totally dependent on the high yields of these few selective cultivated plants. By and large, a dependent, though viable relationship exists among plant domestication, genetic diversity and human population growth.

REGIONS OF CROP PLANT DIVERSITY

A study of the origin of agriculture/cradles of agriculture and the spread of agriculture provides clues to the geographical distribution of centres of plant domestication. These centres would obviously be located in areas where maximum diversity of crop(s) is found. The spread of agriculture could have taken place to more regions of crop plant diversification perhaps the first who indicated about such regions where the initial plant domestication might have taken place. Indicated the cradles of agriculture in a more elaborate manner.

He believed that time is the only factor that influenced the dispersal of a species and its increase of variation. Based on this hypothesis of presence of greatest diversity, he named eight centres where agriculture developed independently. He identified these as 'Centres of Origin' of crop plants. Also, these centres were characterized by the accumulation of dominant genes in the centre and the recessive genes in the periphery.

The criteria used by Vavilov in determining the centre of origin for a crop plant/species were:

- Differentiating plants into specific and infraspecific taxa on morphological and genetic basis.
- Determining the area of distribution of such species and groups of species.
- Establishing the distribution of genetic diversity and determining the geographical centres where this is at its maximum, the centre where diversity of genetically allied species is concentrated, especially those centres with endemic forms/characters.
- Correlating the above distribution/diversity with the areas of concentration of nearest wild relatives.
- Comparing centres of origin of group of cultivated plants with certain specialized parasite(s).
- Support the above, seek linkages/evidences from archaeology, linguistics and history.

However, later, the presence of wild relatives was considered an essentiality in designating a centre as 'a centre of origin'. The designated centres of origin and their boundaries were revised and re-revised subsequently by different authors and more theories and concepts came into focus as more detailed studies and more number of crops/species were taken into account.

NUCLEAR CENTRES AND REGIONS OF DIVERSITY

Definitely, sites of early farming which were discovered through efforts of archaeologists can substantiate the presence of a cradle of agriculture at a place. More or less simultaneously and in different regions of the world.

His concept clearly envisaged centres of agricultural origin from which farming spread into one or more regions for which he proposed the name 'Nuclear centres and regions of diversity'. He linked the nuclear centres with the archaeological evidence to provide strong proofs of agricultural origins.

There are several regions where crops actually did not originate. In these regions, there is paucity or absence of wild progenitors and absence of archaeological remains and such other data, to suggest antiquity of a crop species. These are the regions into which the crops perhaps spread from the nuclear centres in the past and where further spatial isolation in time and intensive human selection played a prominent role in the increase of genetic diversity .

In distinguishing the nuclear centres of agricultural origins from Vavilov's broader 'centres', Hawkes has tried to combine his views with other authors contributing to the knowledge of origin of agriculture and centres of diversity.

Megagene Centres

A close associate of Vavilov, proposed 12 megagene centres of crop-plant diversity. The new areas added to Vavilov's eight centres were Australia, whole Africa and Siberia followed by revision of the boundaries to make 12 centres. Microgene centres of wild growing species related to our crop plants, where the cultigens first originated, were also demarcated.

Zeven and Zhukovsky have dealt this elaborately in the *'Dictionary of cultivated plants and their centres of diversity'*, listing species for different megagene centres, and the range and extent of the distribution of genetic/varietal/specific diversity.

Centres And Non-Centres

Harlan recognised only three main 'Centres', each with a more or less connected but large and diffuse non-centre.

Harlan's centres and non-centres

Centre	Non-centre
North Chinese Centre - B1	South-East Asian and South Pacific non-centre - B2
North East Centre - A1	African non-centre - A2
Meso American Centre - C1	South American non-centre - C2

Harlan also recognised smaller areas/pockets of varietal and/or racial diversity within a Vavilovian Centre, and he termed these as 'Micro centres'. Such small areas, as in Turkey and Africa contain varietal diversity of several crops in the plains and/or mountains.

SPECTRUM OF GENETIC RESOURCES

- Related wild species and weed races in the same genus, and related genera found in the regions of primary or secondary diversity. This class may include undomesticated wild species which are consumed by man as food or for other uses, e.g. forestry species, medicinal plants and pasture species.
- Unimproved landraces (folk varieties) and special purpose types from the areas of diversity which are adapted to specific ecological niches or provide special dietary/religious needs. Inherent diversity is a unique feature of the landraces. Many landraces are varietal mixtures.
- Pure-line selections or open-pollinated commercial varieties from old agricultural areas where production levels remain largely unchanged in the last half century.
- Obsolete varieties which can be found only in germplasm collections. Ecostrains of obsolete cultivars may persist in other areas.
- Advanced cultivars, modern elite varieties (HYVs) and F hybrids developed by scientific breeding and grown in areas of modern intensive agriculture. Composites and synthetics evolved through plant breeding also belong to this class.
- Other products of plant breeding programme or genetical studies, which include breeding lines, breeding stocks, mutants, gene markers, genetic stocks, induced polyploids, aneuploids, intergeneric and interspecific hybrids, and cytoplasmic sources.

These categories also indicate both an evolutionary continuum linking prehistoric wild forms with present day cultivars, and an ecological continuum linking wild and partly domesticated taxa with domesticated forms .

VALUE OF CROP PLANT DIVERSITY

For plant breeders, in their endeavour directed towards increased agricultural production, there is a pressing need for more genetic diversity to work upon, to cater to varied kinds of problems and needs. The wider the range of choice a breeder will have in selecting the appropriate kind of diversity, the better will be the chances for his success for any particular goal. Earlier the breeders were content to go not much further for their material than the old landraces and varieties that were then available in their own countries or from neighbouring ones, but in the last 50 years or so, breeders are requiring a much wider range of genetic diversity. The situation is particularly serious when developed varieties get immediate popularity.

Here, the crops possess a 'narrow genetic base', and uniformity holds a hidden danger for it can offer an open doorway to attacks by pests and pathogens. With a uniform crop variety or a range of rather similar varieties, once a strain of a pathogen becomes adapted to attack it, the whole crop can be lost in a very short time.

In the United States, a single source of cytoplasm that had been used in developing the majority of the corn belt hybrids, also transmitted cornblight susceptibility to all other varieties that contained it. The vulnerability of these maize varieties became evident in 1970 when the advent of southern corn blight led to the widespread destruction of corn crop.

The wealth of landraces, which could not even be called varieties and which the farmers grew earlier, are gradually disappearing. A collection and study of these has revealed that these exhibit many useful characteristics, and can help in crop improvement programmes. Similarly, many wild relatives of crop plants can be useful in present day breeding programmes.

One can thus understand the dire need for conservation of plant genetic diversity. Its value in the future will be much more than what can be imagined at present, considering the diversified crop improvement programmes, technologies and human needs.

THREAT TO GENETIC DIVERSITY

Three processes that affect the crops that are cultivated now, i.e., genetic erosion, genetic vulnerability and genetic wipeout are not mutually exclusive but are, in fact, interlocked by the demands of an increasing human population and rising expectations.

Genetic Erosion

The technological bind of improved varieties is that they eliminate the resource upon which they are based. Over the past 10,000 years, crop plants have proliferated to an innumerable number of locally adapted genotypes. These landraces and folk varieties of indigenous and peasant agriculture have been the genetic reservoir for the plant breeder in crop improvement.

Suddenly this genetic diversity is being replaced with a relatively

Group	Diversity within group	Homogeneity within a strain or population	Agronomic or commercial value	Genetic potential in breeding
Modern elite cultivars	Low to moderate	Very high	Very high	Moderately high
Principal commercial types	Moderately low to moderate	Moderate to high	Moderately high	Moderate
Minor varieties	Moderately high to high	Moderately low to moderate	Moderate	Moderately high
Speciality types	Moderately low to moderate	Moderately high	Moderately low	High
Obsolete types	Moderate to high	Moderately high to high	Moderately low to moderate	Moderately low
Breeding stocks	Moderately low to moderately high	Moderate for lines; low for bulks	Most variable	Moderate to high
Mutants	Moderately low to moderate	Moderately high to high	Mostly low; few moderately high to high	Mostly low
Primitive types	Moderately high to high	Low to moderate	Moderately low	Moderately high to high
Weed races	Moderately high to high	Low to moderately low	Low	Moderate to moderately high
Wild species	Moderately low to moderate	Low to moderately low	Very low	Moderate to high

small number of varieties bred for high yields and other adaptations necessary for high input agriculture. In addition, the scarcity of land is forcing changes in land use and agricultural practices resulting in the disappearance of habitat which harbour the wild progenitors and weedy forms of our basic food plants.

As a result of these two trends, there is urgent need to collect and conserve the diverse genetic materials that remain. In a world where per capita resources are decreasing as the human population grows, the concept of a sustainable future is becoming increasingly more important. Biological diversity is one of the components of any sustainable future that includes human beings. Safeguarding this resource through scientific management should be our goal.

Genetic Vulnerability

Genetic vulnerability is the risk of high input agriculture with commercial food crops varieties typical of developed nations. Genetic erosion, the gradual persistent loss of plant genetic diversity, is most typically but not exclusively, a phenomenon of landraces in developing nations.

Genetic vulnerability is the 'thin ice' of a narrow genetic base. Never before have there been such widespread monocultures (dense, uniform stands of billions of plants) covering thousands of acres, all genetically similar. The narrowness of the genetic base is responsible for greater risk of crop failure as occurred in the wheat stem rust of 1954 or the southern corn blight of 1970 in the US. The Irish potato famine in the 1840s is a classic example of genetic vulnerability.

Genetic Wipeout

The third threat to crop plant germplasm is the rapid and wholesale destruction of a wealth of potential species constituting genetic resources. Social disruptions/instability can eliminate such promising diversity.

The genetic heritage of a millenium in a particular valley can disappear in a single bowl of porridge if the seeds are cooked and eaten instead of saved as seed stock. Equally dramatic is the discarding of a genetic collection because a curator retires or the collection is no longer of use to the institution.

These delicate points need to be taken care of, should such eventualities arise and germplasm fully safeguarded. Institutional arrangements by which the perpetuation of genetic collections can be coordinated nationally, regionally and internationally need to be further streamlined.

CONSERVATION OF GENETIC DIVERSITY

Genetic conservation programmes are directed towards the long-term preservation of genetic resources either *in-situ* or *ex-situ* so that the potential for continuing evolution or improvement would be sustained. Thus, conservation is more inclusive than preservation, the latter provides only for maintenance but not for evolutionary modifications under different environments .

It is impractical to conserve all the available gene resources and the primary goal should be to conserve as many representative samples of existing germplasm as human resources permit. Priority should be given to those being threatened by extinction or displacement and genetic erosion. The targets of conservation may be one or more of the following:

- A nucleotide pair - a mutation.
- A desirable or favourable allele of a gene.
- A gene - complex controlling a desirable trait.
- A co-adapted gene complex.
- A chromosomal segment and.
- A cytoplasmic component or components. Genetic conservation partly covers the known genes and partly aims to save plant materials of yet unknown genetic potential, such as the populations of threatened landraces/rare or endemic taxa of wild relatives. The latter category must necessarily be the primary responsibility of the conservationists because of the large number of species included. Germplasm is non-replaceable and its protection must be safeguarded.

Conservation Systems/Programmes

A comprehensive genetic conservation programme for a given crop using *ex-situ* preservation should include the following components:

- Survey.
- Acquisition/exploration and collection.
- Maintenance/multiplication/rejuvenation.
- Evaluation.
- Documentation.
- Distribution/exchange.
- Preservation.
- Training.
- Collaborative network. Plant quarantine facilities and measures should form part of a large genetic resources centre.

Conservation In-Situ

- Protection of vast tracts to conserve plants and animals in entire biomes - extinction of species is deterred, but this has little impact on useful plants.

Wild Species in Natural Communities

- Domesticates - landraces in their areas of cultivation.

Conservation Ex-Situ

- Domesticates in mass reservoirs or genetic reserves under the dynamic type.
- Forest reserves (provenances) in the wild.
- Conservation of seeds, plants, plant parts, cells, tissues and meristem cultures under a static environment (genebanks).

Botanical gardens or parks also belong to this category as the last line of defence, but they also generate useful information and provide educational benefit.

Seed conservation is by far the most efficient means of genetic conservation. Seed storage in genebanks may be organised on the basis of projected seed longevity and seed rejuvenation cycle. Two types of genebanks are designated by the FAO Panel of Experts

Base Collections

Seeds of 5 - 6 per cent moisture content are sealed and stored between -10 to -20°C for long-term conservation; the scope is generally comprehensive; a centre holding such a collection assumes responsibility for viability tests records and other associated matters. The seeds are distributed not for use but for rejuvenation . A duplicate site is needed for security.

Active Collections

Seeds are dried to 8 ± 1 per cent moisture content, sealed and stored in medium-term storage (slightly above 0°C), from which samples are drawn for regeneration, multiplication and distribution, evaluation and documentation. Active collections complement the base collection and are often associated with plant breeding or plant quarantine stations. Records on origin, handling and distribution of accessions should be maintained.

Working collections of plant breeders and other disciplines are not considered part of the genetic conservation system. Collections of parents and other breeding materials are generally held in short-term storage and their maintenance is subject to the whims and fancies of the breeder .

The evaluation data should be made available to other interested workers for use in breeding programmes. Therefore, promising breeding materials, such as F_1 hybrids, F_2 and F_3 bulks, composite crosses, etc. should also be carefully stored for a period of time in order to enable other breeders to be able to have access to the materials. The importance of plant genetic resources as basic materials for crop improvement has been highlighted. The circumstances leading to settled agriculture, and the dynamics of plant domestication resulting into changes in plants from wild to cultivated. A broad picture has been presented on the centres of origin/diversity in crop-plants and recent views on this topic are expressed.

The spectrum of genetic diversity covering different categories of genetic resources is indicated. The importance of crop plant diversity for increased food production is stressed, both in terms of its collection and conservation. The concern on genetic uniformity and genetic vulnerability *vis-a-vis* genetic erosion has been emphasized.

Chapter 7

Biotechnology and Resources Conservation

Biotechnology has brought about a revolution in the way that plant genetic resources can be utilised. No longer are breeders confined to seeking genes useful for improvement among the closest relatives of a target genotype. Wide crossing can be facilitated by embryo rescue and potentially by protoplast fusion. Selective gene transfer by somatic cell genetic techniques ('genetic engineering') can, in theory, transcend the widest taxonomic barriers.

Related molecular technologies can aid disease indexing and germplasm characterization. At the more prosaic level, cloning *in-vitro* and the exchange of *in-vitro* plantlets make it possible to multiply and distribute elite genotypes with a speed, convenience and security far in excess of traditional technologies. Meristem-tip culture and shoot-tip grafting alone or combined with other forms of therapy can free plants of viral pathogens. This is particularly important for clonal crops that lack the 'filter' of seed production .

These advances in biotechnology, brought about by a combination of *in-vitro* (tissue culture) technology, genetics and molecular biology, are changing the way that we look at genepools and at the potential for crop improvement. The most dramatic benefits may well be reaped in the case of crops that are sterile and/or are traditionally propagated vegetatively.

Genetic engineering can bring about improvements in their hitherto virtually inaccessible genomes and the vegetative habit, amplified by *in-vitro* cloning, stabilizes the improved genotype.

Any resource of value clearly needs to be protected and plant genetic resources are no exception. Awareness of the value of genetic resources has been sharpened in recent years by threats to biodiversity in an often hostile world. This realization has come at the same time as the realization that new biotechnologies can greatly increase our ability to use genetic resources.

For crops that produce seeds capable of storage under conditions of low temperature and low humidity ('orthodox seeds') their genetic conservation can be managed very adequately by seed genebanks. However, this technical approach requires access to sufficient quantities of storable seeds. Significant amount of genetic resources fall outside that definition.

At one end of the spectrum are those materials that are not in cultivation or that have not been targetted for conservation efforts for other reasons. They are to be found in the natural environment and are probably best conserved there *'in-situ'*. At the other end of the spectrum are cultivated genotypes and their relatives that are recognised as being in need of *'ex-situ'* conservation but that do not respond to conventional seed storage methodology.

These problem crops fall into two categories: 'recalcitrant' seed producers and clonal crops. Recalcitrant seeds die when exposed to desiccating conditions or low temperatures. This presents serious problems to the breeder in both the short and long term. Seeds cannot be saved either from one generation of the crop to the next, nor can they be saved for future human generations. Examples of recalcitrant seed-producing species are mango, cacao, coconut, walnut, jackfruit and avocado.

Clonal crops cover a wide range of species from the root and tuber crops, such as potato, cassava, yam, taro and sweet potato to fruits, such as apple, pear, citrus, banana and the cooking banana (plantain). Other miscellaneous crops including vanilla, ginger, turmeric, hops and sugarcane are also clonally propagated.

In some of these cases, seed production is impossible due to sterility. In others, it is undesirable to produce seeds for conservation as this would break up highly heterozygous clonal genotypes. A significant number of the problem species are important in South and South-East Asia in particular, and the developing world, in general.

For both types of problem crops, conservation needs are met, traditionally by the use of the field genebank ('orchard', 'clonal repository', etc.). For reasons of security, cost, plant health and convenience, the field genebank has many disadvantages. Its advantages are few and relate to the opportunities that are offered for evaluation of the germplasm. However, this is not a function of a base genebank. In effect, recalcitrant seeds and clonal crops do not have, at present, a satisfactory, safe, base conservation mode.

In the rectification of this deficiency, we find another role for biotechnology in addition to those identified above. If genotypes can be inoculated into *in-vitro* culture for purposes of clonal propagation and germplasm distribution, why not prolong the *in-vitro* phase and use it for storage? The realization of the potential of *in-vitro* conservation came in the early 1970s at which time the storage of microbial cultures was virtually routine and the storage of animal tissue cultures a practical proposition, if not entirely routine.

In the last 20 years, we have seen dramatic advances in protoplast culture methodologies, in genetic transformation technology and in the extension of *in-vitro* propagation to most species that have been attempted. Over the same time scale, *in-vitro* conservation has moved from speculation to development and implementation. An explanation of the technologies involved in this evolution. However, it is relevant firstly to point out a further linkage between *in-vitro* conservation and biotechnology.

The different aspects of biotechnology involve the utilisation of a range of *in-vitro* culture systems: protoplasts, cells, callus, embryos, shoots and plantlets. Often, these are by their very nature ephemeral, highly dependent on external inputs for their continued health and growth and, in some cases, genetically unstable.

The phenomenon of 'somaclonal variation' is a double-edged sword that offers one way of generating variability *in-vitro* but that also threatens the stability of genotypes passing through the more susceptible unorganized culture systems. Accordingly, *in-vitro* cultures, the raw materials of plant biotechnology, have a conservation problem of their own. This has several aspects.

Firstly, there is a need to arrest time-related changes including somaclonal variation. Then there is a need in routine *in-vitro* culture work to store stock cultures in an economical and safe way, free of contamination and other risks. In experimental work, there is a need to 'suspend time' to enable direct comparisons to be made between samples processed over a period.

Finally, many of the genotypes evolved *in-vitro* do not have a natural storage form. Valuable but unstable callus cultures used in secondary product synthesis *in-vitro* may be incapable of conversion to a whole plant, having lost their 'totipotency'. Therefore, it follows that they cannot form seeds. In this way, the materials of plant technology themselves form a third problem category to add to the recalcitrant seed-producing and clonal crops.

In-vitro Conservation: Foundation Technologies

Although the essential core of plant genetic resources conservation is storage *per se,* the term 'conservation' covers the entire process from collecting through to access for evaluation and utilisation. This breadth of definition, encompassing a 'package' of technologies is true also for *in-vitro* conservation. The foundation technologies that make up an *in-vitro* conservation system are collecting, disease eradication and indexing, culture initiation, multiplication, storage and distribution.

As in conventional conservation, there are points where monitoring should take place to check viability and genetic stability as well as the important tasks of characterization and documentation.

Orthodox seed storage is carried out using methodologies that offer either 'active' or 'base' conservation conditions for the medium-term or long-term respectively and with different criteria in relation to access, study, multiplication, utilisation and distribution.

In parallel, *in-vitro* conservation has active and base components, which gives it an advantage over the field genebank which, as indicated earlier, is an active conservation mode. *in-vitro* active conservation is achieved by maintenance of cultures in the growing state, usually with growth slowed down by limiting physical environmental or culture medium factors.

In-vitro base conservation is achieved by cryopreservation, i.e., storage in liquid nitrogen at or near a temperature of -196° C .The linkages between the different foundation technologies of *in-vitro* conservation and their relationship to the field genebank. It is clear that *in-vitro* conservation is more complex but, by the same token, has more to offer than conventional conservation.

With the exception of some examples in biotechnology involving, non-totipotent callus cultures as detailed earlier, culture initiation, multiplication and plant regeneration form the cycle of events that is the backbone of an *in-vitro* conservation scheme. The way in which this cycle is carried out has implications for the success of the conservation effort. There are two main areas where this is important.

The first is in relation to genetic stability. As already indicated, certain culture systems are less stable than others, being more prone to somaclonal variation. Therefore, where there is a choice, *in-vitro* conservation of plant genetic resources generally avoids the use of cell and callus cultures. (The use of protoplast cultures would be difficult to justify in any conservation context).

The remaining culture systems, embryos, shoots and plantlets are all more suitable. Until recently, there has been a strong prejudice favouring the use of shoots/shoot-tips/meristems for cryopreservation and shoots/plantlets for slow growth storage. While the latter choice can still be justified, there is an accumulation of evidence and experience that would say that embryo cultures have advantages in cryopreservation. These largely relate to the second area of concern in the *in-vitro* culture cycle, i.e., amenability to storage. Structural factors, particularly size and degree of complexity, strongly influence the reaction of a culture system to cryopreservation. In general, the smaller and less complex the structure, the more likely it is to survive cryopreservation. This, unfortunately, is virtually the opposite of the criteria for judging the likelihood of somaclonal variation. A compromise may, however, be found in the somatic embryo (or indeed, the immature zygotic embryo) which may well combine a small unit size with a low risk of somaclonal variation.

The point of entry of germplasm into the *in-vitro* conservation cycle may vary. Sometimes, the germplasm will be transferred in the form of a culture from another *in-vitro* collection. Sometimes it will be inoculated *de novo* into culture in the *in-vitro* genebank. There is a third option, that of *in-vitro* collecting. This involves the introduction of material into culture actually in the field. Explores this novel and very useful yet simple application of an *in-vitro* technique in the service of genetic conservation.

In-vitro Collecting

Plant explorers have had many decades and more of experience in collecting germplasm from its natural environment or area of cultivation and transferring it to other locations and to genebanks. However, even the most experienced collector faces problems that threaten the efficiency of the collecting mission and the security of the germplasm collected. These problems waste resources and lead to inadequate sampling of target areas and genepools.

Even with good planning, it is possible to find insufficient supplies of seed or other suitable propagules or to find that the seed is immature. These problems may be due to seasonal variations or browsing by animals among other causes. In the case of vegetative propagules such as would be collected for say, banana, the propagule is prone to deterioration within a short period of time. This limits the duration of the collecting mission that can be undertaken and the area of territory that can be penetrated without complex transport arrangements for personnel and collected material.

Recalcitrant seeds are similarly prone to deterioration and have a second problem in some cases such as coconut, of exessive weight/ bulk. A crop such as cacao combines virtually all the problems, the two types of material normally collected being budwood that is prone to fungal attack in transit or large fruits containing recalcitrant seeds.

Over the last 8 years, IBPGR has been sponsoring research to develop an alternative, *in-vitro* based approach to collecting plant germplasm in the field. This involves the use of minimal equipment, modified surface sterilization procedures to reduce toxicity and modified culture media, sometimes containing antibiotics and/or fungicides to control residual contaminants. Using these basic principles, methods have now been researched for collecting materials as diverse as cacao, coconut, forage grasses, temperate fruits and citrus.

In-vitro collecting is an approach rather than a rigid set of prescribed procedures. It needs and, indeed, offers to the collector great flexibility to adapt the precise methodology to his own situation. An example to illustrate the flexibility of the technique can be found in coconut.

Coconut embryo culture is well developed providing the foundation for a field collecting method that avoids the transportation of large quantities of heavy nuts. Several scientists have tackled the adaptation of embryo inoculation method using different approaches. In one case, considerable efforts have been made to reproduce the conditions of cleanliness found in the laboratory by using an inflatable glove box in which the standard sterlization, dissection and inoculation procedure can be carried out.

Rather less sophisticated is a second approach wherein plugs of endosperm containing the embryo are extracted in the field using a cork borer, surface sterilized and then dissected to reveal the embryos which are inoculated into culture. The only protection from contamination is afforded by a wind shield, a spirit lamp to flame instruments and necks of vessels and importantly, the dexterity of the scientist.

A further stage in the simplification of the procedure is to defer dissection and inoculation until the material reaches the laboratory. In this case, the sterilized endosperm plugs are inoculated into a sterile saline solution for transportation). The endosperm plugs are, if necessary, resterilized in the laboratory and dissected, the embryos then being transferred to normal medium.

Finally, we find a method that completely excludes any complex field procedure. The nuts are simply dehusked and cracked open in the field. The endosperm plugs are extracted and transferred to a plastic bag containing coconut water. The bag is closed and transferred to an insulated container for transport to a laboratory for sterilization, extraction and inoculation of the embryo in a laminar air-flow cabinet.

Efforts to test *in-vitro* collecting in a wide range of circumstances and with collector's new to in-*vitro* manipulations have been largely successful. Examples of field testing include collecting coconut in Indonesia and forage grasses in East Africa, in a collaboration between

IBPGR and the International Livestock Centre for Africa. In 1990, IBPGR and CATIE in Costa Rica held a training course on *in-vitro* collecting supplementary collecting method alongside traditional approaches. It is a good example of appropriate biotechnology helping conservation.

In-vitro Storage: Principles, Successes and Continuing Challenges

There are two basic options for *in-vitro* storage, slow growth for the short-to-medium term and cryopreservation for the long-term (Since the definition of storage time-span and the concepts of active or base storage derive strictly from motivation rather than methodology, there is no reason why, in fact, cryopreservation should not have applications in short-to-medium term conservation.

The principle of slow growth storage is that the safety of *in-vitro* culture be used without disadvantages of frequent subculturing. Thus, the inputs in terms of labour and consumables are reduced, as also the risks of contamination at each transfer interval. However, the cultures can be observed to be growing and can readily be restored to normal culture conditions for multiplication and distribution to users.

The subculturing interval can, in a few fortunate cases, be extended without significant modification of culture procedures. Generally, it requires culturing at a reduced temperature or in the presence of retardants.

The disadvantages of slow growth are few but relatively important. The stress imposed on a culture to retard its growth will, inevitably, affect its vitality. A careful balance needs to be maintained to avoid general loss of viability and, worse, selective loss of viability that could lead to genetic drift. This might seem to be a strange concept in work with putative clones.

However, it is probably safer to consider *in-vitro* collections as 'tight populations' rather than absolute clones. This is certainly the case for unorganized cultures, such as callus, and may be true for some organized material. For these reasons, whilst slow growth storage does reduce the direct inputs to culture maintenance, it still gives a burden of monitoring efforts to the curators of the culture collection.

Despite these cautions, we can find an impressive number of notable successes in the routine maintenance of germplasm in the form of shoot cultures or plantlets in slow growth. Significant examples can be found within the IARCs for root and tuber crop germplasm, the International Potato Centre (CIP) maintaining collections of potato and sweet potato, the Centro Internacional de Agricultura Tropical (CIAT), cassava; and the International Institute of Tropical Agriculture (IITA), yam and cassava. Other examples include the banana and plantain collection maintained at the INIBAP Transit Centre in Leuven, Belgium, and numerous collections of clonal crops conserved by national programmes and research laboratories around the world.

The continuing challenges with slow growth storage fall into four categories. Firstly, we do not know the breadth of applicability of the technique in terms of genotypes within a species to which it is suitable or among species.

As with the development of *in-vitro* propagation technology, it will, for the most part, be a case of trial and error with only limited scope for anticipating reactions and needs of particular genotypes. This will all take time and individual efforts on the part of those wishing to apply the technique.

Secondly, there has not in any one case been an absolutely full evaluation of the interactions of different environmental factors (basic medium composition; retardants, temperature, light level, quality and photoperiod). Therefore, we know little about the latitude for varying one or more of these factors. This makes technology transfer potentially problematic, since not all the factors are documented even if appreciated.

Thirdly, we do not know the maximum period of time for which slow growth can safely be applied. There is a reasonable amount of information available on the duration of the typical subculturing interval but not on how many can be applied sequentially or whether the interval remains constant over time. There are anecdotal suggestions of cultures adapting ('habituating') to slow growth conditions.

Finally, there is a knowledge gap in relation to genetic stability or even phenotypic stability in material exposed to slow growth conditions for extended periods of time. There is a limited amount of information available from studies of callus cultures to suggest that there may either be selection or persistent physiological change brought about by slow growth at reduced temperatures.

The same observation has recently been made for callus stored under mineral oil overlay. Evidence on organized cultures is lacking. Accepting that we should expect them to be more stable, it is still necessary to apply a degree of caution in recommending their storage in slow growth without reservation for extended periods of time.

In summary, the challenges with slow growth storage are to press on with applying it as widely as possible, learning about its scope and limitations, and gathering data to build up a picture of its effects upon cultures in all respects.

Cryopreservation is a very different phenomenon from slow growth with its own advantages, disadvantages and challenges. Because all metabolism is suspended at the temperature of liquid nitrogen (- 196° C) or, as is often used for storage, liquid nitrogen vapour (- 150° C), time is not a critical factor in the design and application of cryopreservation regimes.

Provided that stable storage conditions be maintained, the only known cause of damage during cryopreservation is free radical activity. This may be initiated by endogenous reactions or may be initiated by background radiation. In either case, some protection can be afforded by chemicals such as dimethyl sulphoxide (DMSO) which are 'radioprotectant' as well as cryoprotectant.

Rather than time in storage being an important factor in cryopreservation, the most critical stages are the transition to and from the frozen state. With the exceptions of some extremely cold tolerant/ freeze tolerant materials, this is a highly unnatural transition and measures must be taken to protect both cellular and tissue structure and the metabolic processes against deleterious effects. These measures can and need be applied at each stage of cryopreservation process: pregrowth, cryoprotection, cooling, warming, the immediate post-thaw period and recovery growth.

There is now great accumulation of information on the variables applying to each stage. In the case of cell suspension cultures, a virtually routine cryopreservation method is available (Withers and King, 1980). For other systems, this is not the case. A summary of the various methodologies attempted is not easy to make. There are great differences between culture systems and also interactions within and between the various stages. Thus, a particular cryoprotectant or recovery medium may be more effective at one cooling rate than another. However, some general observations can be made concerning the progress that has been made in elucidating different stages and what remains to be researched.

'Pregrowth' refers to the period of time (usually measured in days) in culture before cryopreservation proper begins. It offers opportunities to induce physiological changes in the specimen to increase its freeze-tolerance. Most effective pregrowth treatments involve some type of 'stress conditioning' (high osmoticum, low temperature etc.).

Among the most interesting observations made in recent years regarding pregrowth are the finding that the use of high levels of sucrose (0.75M) for somatic embryos of oilpalm can eliminate the need for cryoprotectants and reduce dependence on precisely 'tuned' cooling rates and that cold hardening *in-vitro* combined with controlled desiccation can mimic the *in-vitro* hardening of fruit buds enabling the cultures to be frozen, again without cryoprotectants.

Cryoprotection is the phase of minutes to hours during which single protectant compounds or several-component mixtures are applied. The most commonly used cryoprotectant is DMSO which can also act as an effective pregrowth additive. DMSO can be used alone for organized cultures. It is more frequently used in mixtures, e.g. with glycerol and sucrose for unorganized cultures .

Sucrose is a naturally occurring compound in plants arid a little research has been carried out in the use of compounds such as proline that are known to be induced under conditions of stress in nature. However, in general, there would appear to be great scope for exploring the potential of natural cryoprotectants, either induced by pregrowth stress conditions such as outlined above or exogenously applied. One of the attractions of using more natural compounds is that it reduces the risk of cryoprotectant toxicity, a problem that can be encountered with high concentrations of DMSO and with old or impure stocks of the chemical.

Cryoprotection is followed by cooling. This can be carried out in a number of ways depending upon the mechanism of survival in use. The mechanism also has implications for the concentrations and types of cryoprotectants used. Two basic mechanisms exist. The first and most commonly used is that of protective dehydration. Slow cooling will cause extracellular ice formation which leads to the extraction of water from the cell and a reduction in the amount of potentially damaging ice formation when the cell contents eventually solidify.

There can be a fine balance between under and over dehydration. It is important to find this as overdehydration can lead to 'solute damage' and excessive shrinkage of the cell with effects such as the loss of membrane surface area causing membrane rupture upon re-expansion. There are experimental procedures to arrive at the optimum amount of dehydration.

In some slow cooling regimes, it is beneficial to remove the superficial water surrounding the specimen by blotting using a filter paper. It is even possible to carry out a certain amount of dehydration in an air stream or in a desiccator before cooling.

The alternative to slow cooling is rapid cooling which avoids ice damage through the induction of either microscopically small ice crystals or avoids ice formation totally through the process of 'vitrification'. Expedients such as direct exposure to liquid nitrogen have been used to achieve survival in specimens treated with conventional, relatively low concentrations of cryoprotectants.

However, vitrification uses extremely high concentrations that are potentially toxic at non-frozen temperatures. Although they facilitate beneficial solidification of the specimen into an ice-free mass, they can cause damage between the time of addition and removal. Therefore, a lot of attention is given in vitrification work to the processes termed 'loading' and 'unloading'.

Vitrification has proved highly successful for animal embryos. It has been applied to various *in-vitro* culture systems from protoplasts to shoot-tips. In the latter case, there is the greatest potential benefit since, as indicated below, large and highly organized specimens tend to suffer structural injury in dehydration-based cryopreservation procedures.

Warming to achieve thawing is simply carried out but is a critical stage in the cryopreservation process. Where there is a risk of ice recrystallization as in rapidly cooled specimens, it is essential that thawing be carried out rapidly as well. There are some examples of successful, even beneficial slow thawing after slow cooling but, in general, it is wise to use rapid thawing as a first option.

The freshly thawed specimen is extremely vulnerable. In the literature, we can find examples of sensitivity to rehydration and plasmolysis injury manifested as weakness in the cell membranes. There is evidence for the infliction of metabolic lesions including respiratory stress. Cell division is delayed and structural injury is visible at gross and microscopical/electron microscopical levels, with corroborative biochemical data indicating membrane breakdown.

Sensitivity to high light levels is detected in some cases, probably indicating photoxidation involving free-radical mechanisms. The frozen and thawed cell is likely to have suffered metabolic insults that lead to the disruption of the normally efficient free-radical scavenging mechanisms in the cell.

There is some evidence that the provision of free radical scavengers can be beneficial during the early stages of recovery. In general, it is wise to take all measures to stabilize the specimen at this stage by incubating under low light levels, by avoiding osmotic disturbance (e.g. by washing) and by using semi-solid rather than liquid medium. If it is felt necessary to remove residual cryoprotectants, this can be done by draining on a filter paper and/or moving the specimen, carried on a filter paper, through a series of dishes of fresh medium to allow the cryoprotectant to diffuse away.

The recovery stage involves the resumption of normal metabolic processes, the resumption of cell division and the further growth and/ or development of the specimen. In the case of callus cultures and cell suspensions, this is a relatively simple process involving the accumulation of a mass of undifferentiated cells.

In the case of organized cultures, it is more complex and involves the re-initiation of meristematic activity. If the specimen has been severely damaged in freezing and thawing, this may be problematic. It is a common observation that shoot-tips undergo a period of undifferentiated growth through callus formation either alongside development of the original shoot meristem or in place of it.

In the latter situation, there is a serious risk of somaclonal variation, in the former case the risk is less but not to be ignored. The use of growth regulators to stimulate growth after freezing and thawing can exacerbate this problem and the use of auxins in particular is best avoided.

The problems of significant damage in both quantitative and qualitative terms encountered in the cryopreservation of organized structures is probably the main motivation for exploring the use of vitrification. However, the scope for modifying the conventional slow cooling-based cryopreservation method should not be underestimated. Systematic experimentation to optimise the variables at the pregrowth, cryoprotection, cooling and recovery stages can yield great benefits .

Vitrification is one of the two most interesting new developments of recent years in cryopreservation; the other is the adaptation of artificial seed technology. Artificial seeds have been explored as a means of supporting somatic embryos, both nutritionally and structurally, of transporting and delivering them to give a degree of convenience approaching that of natural or pelletted seeds. Success in these endeavours has been significant, although there is much research to do to make the technology universally applicable and economic.

One of the exciting spin-offs of this research may be an improvement in cryopreservation technology for organized cultures such as somatic embryos and shoot-tips, taking advantage of research into the desiccation process as well as the encapsulation itself. There is already evidence that the encapsulation of shoot-tips is a practical approach to cryopreservation that may lead to improved survival and maintenance of organization. This is an area of research to follow closely, particularly as it can take advantage of the attractive somatic embryogenesis propagation system.

In-vitro Conservation of key Crops

An evaluation of the current status of *in-vitro* conservation can be gained by reviewing the progress that has been achieved in creating the 'package' of technologies that complete an *in-vitro* conservation scheme for a given crop. The target crops for *in-vitro* conservation are many and diverse. Three instructive examples to review in this way are coconut, banana/plantain and cassava. They are all important staple crops in developing countries and have all received a significant amount of attention from the point of view of *in-vitro* culture.

In the context of *in-vitro* conservation, coconut is important as a recalcitrant seed-producing crop. It has received a lot of attention from *in-vitro* culture technologists because it is impossible to clone by conventional means. The clonal propagation *in-vitro* of oilpalm has become routine in recent years. Despite some concerns over genetic stability in certain propagation systems, the advantages of being able to clone oilpalm are tremendous and this has served as a model to encourage efforts on coconut. When we look at the state of the package of *in-vitro* conservation technologies for coconut, however, these efforts are not yet reflected in successes.

In-vitro collecting is highly developed and the same technology can be used to exchange germplasm between genebanks. Unfortunately, though the culture system used involves one-to-one regeneration from zygotic embryos, this severely limits the realization of the potential of both *in-vitro* cloning and *in-vitro* storage.

Technically, cultured plantlets germinated from zygotic embryos could be subjected to slow growth storage but the only advantage that this would offer over the field genebank would be potential protection from infection with pathogens. Without extensive investigation to optimize slow growth conditions for a sufficiently wide range of genotypes, it would be an extremely risky conservation mode since it lacks the opportunity to replicate valuable genotypes in culture.

Cryopreservation of mature zygotic embryos has been achieved with some very limited success. Early indications from IBPGR-sponsored research carried out by ORSTOM/IRHO in France and Cote d'Ivoire are that immature embryos of coconut may be responsive to Cryopreservation. However, the problem remains of applying storage to unreplicated culture units.

The lack of a clonal propagation technique is an extremely serious bottleneck in the conservation and utilisation of coconut germplasm by biotechnological means. Other lacunae in the case of coconut exist in the area of disease indexing, disease eradication and quarantine. There may be scope for some *in-vitro* approaches here, particularly in providing a contained quarantine method.

The biotechnological modification of coconut genotypes is virtually unexplored and will again meet a bottleneck in the area of clonal propagation.

Bananas and plantains present a different type of challenge from coconuts. Some but not all genotypes produce seeds. The storage of seeds has been found difficult but they are not recalcitrant. Vegetative propagation is the normal method of multiplication and clones are the main target for conservation efforts. In respect of *in-vitro* techniques, a much more satisfactory situation exists for banana and plantain, where clonal propagation is virtually routine, *in-vitro* collecting has been attempted with some success but not widely tested, and *in-vitro* exchange and slow growth are used in routine.

The INIBAP Transit Centre in Leuven, Belgium among others exchanges cultures and maintains a large collection in slow growth. Meristem-tip culture has had some success in eliminating viruses from shoot-tips in this crop but is not entirely satisfactory due to the large size of the meristem compared with other examples such as potato and cassava.

Success in attempts to cryopreserve embryogenic suspension cultures of a cooking banana cultivar (Bluggoe) has been reported recently. The Cryopreservation method is based on cryoprotection with DMSO and slow cooling. Plant regeneration has been achieved through somatic embryogenesis. Earlier attempts to cryopreserve the large and fleshy shoot-tip of banana were totally unsuccessful, illustrating the advantages that can be gained by the use of embryogenic systems.

However, a major problem with bananas and plantains relates to genetic stability *in-vitro.* Some genotypes in particular appear to be prone to high levels of instability that seriously affect field performance . There is a need to determine the causes of this instability to understand better its linkage to genotype, to develop methods for detecting instability at a sufficiently early stage in the propagation/storage cycle, and to determine whether storage conditions affect the incidence of instability either positively or negatively.

The improvement of banana and plantain genotypes by using biotechnological approaches must be considered within reach. The range of culture systems available in the genus is sufficient to support genetic manipulation studies and there are already reports of the successful isolation of somaclonal variants with fusarium wilt resistance. The disease aspects of coconut are equally applicable to banana and plantain.

Let us look now at the final model chosen for this analysis. Cassava is normally propagated vegetatively and most conservation efforts are devoted to the storage of clones, despite the fact that seeds can be produced and stored . Cassava has been field collected using an *in vitro* approach (Roca, personal communication). It is readily introduced into culture and propagated clonally via shoot nodal cuttings and can be induced to form somatic embryos. Slow growth storage is carried out in routine. IBPGR and CIAT are currently finalizing a project to test management standards for *in-vitro* active genebanks using cassava as a model.

Cryopreservation of cassava seeds and zygotic embryos is feasible and several scientists have investigated the possibility of cryopreserving *in-vitro* cultures. It is now possible to achieve survival and organised recovery at a high frequency in the case of shoot-tips, and somatic embryos are also found responsive. There is now the foundation for developing an *in-vitro* base genebank for cassava. *In-vitro* exchange is routine for this crop and, in combination with meristem-tip culture and associated techniques to eradicate virus, makes a major contribution to the health status of cassava germplasm in circulation and conservation.

Efforts are currently under way to apply genetic manipulation techniques to cassava with a view, in the first instance, to transferring pest and disease resistance. Therefore, of the three models, cassava is having the closest fully feasible and operational *in-vitro* conservation and utilisation package. While comparison of the situation for cassava with that of coconut might give cause for disappointment in the case of the latter crop, it should be remembered that it is only very recently that the picture was so optimistic for cassava.

Once the major bottleneck of *in-vitro* clonal propagation has been solved for coconut, the path will be clear for developing the other components of the package.

Complementary Conservation Strategies

Rather than concentrating thoughts on technologies, the starting point for devising a conservation strategy should be the genepool. The needs of a genepool are very unlikely to be met by a single conservation technology. While it is not always the case in practice, it is eminently logical to have a balanced application of both *in-situ* and *ex-situ* approaches to conservation. Within the latter category, a further balance needs to be struck between seed, field genebank, *in-vitro*, pollen and, perhaps in the future, DNA storage.

Each technology should be chosen on the basis of utility, security and complementarity to the other components of strategy. The inappropriate application of sophisticated technologies, such as the unnecessary use of *in-vitro* cloning and storage, or its use without a sufficiently rigorous experimental foundation is highly inadvisable. It may endanger germplasm through storing it under suboptimal conditions or with an inadequately wide genetic base, it will damage the reputation of biotechnology, and will certainly waste financial resources.

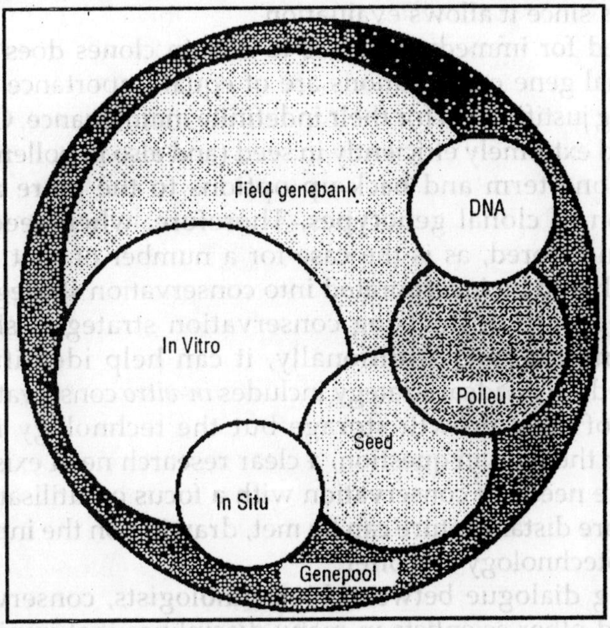

Fig. A Hypothetical Model Showing the Complementary Use of Different Conservation Technologies

This caution applies particulary to the use of molecular technologies. They have a number of very valid applications in the improvement of genotypes by selective gene transfer, in increasing our knowledge of genetic diversity by the application of Restriction Fragment Length Polymorphism (RFLP) and other associated technologies alongside isozyme analysis and conventional morphological characterization, and improvement of disease diagnostics by the development of broad and narrow spectrum probes.

However, whilst some specific niches can be foreseen for the use of DNA storage within a complementary conservation strategy, it is difficult to see this approach safely overtaking the other available conservation technologies.

For the problem crops considered here, a significant part of the genepool will certainly be conserved either in the field genebank or *in-vitro* for the foreseeable future. The two approaches are truly complementary, each balancing the advantages and disadvantages of the other. The security of *in-vitro* conservation and independence of climatic conditions balance the risks and climatic specificity of the field genebank. *In-vitro* conservation is also likely to be less expensive. However, the field genebank is an important adjunct to improvement programmes since it allows evaluation.

The need for immediate access to certain clones does not mean that all clonal gene combinations are of equal importance over time with a strong justification for their indefinite maintenance. Genes may be conserved extremely efficiently in seed (and also in pollen), offering important long-term and back-up options to the more accessible conservation of clonal genotypes. Therefore, when seeds can be produced and stored, as is the case for a number of root and tuber crops, this should be incorporated into conservation strategies.

This approach of devising conservation strategies should aid efficient conservation. Additionally, it can help identification of research needs. If an ideal strategy includes *in-vitro* conservation under conditions of slow growth storage but the technology is not yet available for the crop in question, a clear research need exists. In this way, the true needs of conservation with a focus on utilisation in the near and more distant future can be met, drawing on the inputs of the best that biotechnology can offer.

A strong dialogue between biotechnologists, conservationists, breeders and other scientists in many disciplines that impinge upon genetic resources conservation and use is the best way of ensuring the necessary stimulus to research, its adaptation to pressing problems and the realization of new bio-technological solutions in agriculture.

Efforts to develop conservation strategies that incorporate biotechnological approaches have been made by the International Agricultural Centres of the Consultative Group on International Agricultural Research (CGIAR) along with their partners in National Programmes. Some examples of the contributions already made to the overall global effort by the IARCs.

IBPGR collaborates with many partners in such efforts. In research itself, IBPGR makes efforts through international crop genetic resources networks to bring together users and curators of crop germplasm. They, collectively with scientists doing research on the technologies necessary to store, document and utilise the germplasm, and have the necessary expertise to develop priorities for action to enhance the safe conservation and utilisation of crop genepools.

Biotechnology has applications in the utilisation of plant genetic resources through mass clonal propagation *in-vitro,* in genetic manipulation, in aspects of phytopathology and safe distribution of germplasm, in characterization and, importantly, in conserving germplasm *in-vitro* to reinforce the benefits of the other applications. *In-vitro* conservation is particularly valuable for crops that produce recalcitrant seeds and those that are clonally propagated.

Their conventional conservation in the field genebank is hazardous in several respects. *In-vitro* conservation is a package of technologies commencing with collecting and leading through disease indexing, disease eradication and storage eventually to distribution and utilisation. *In-vitro* collecting is useful in overcoming problems relating to the availability, condition and portability of germplasm to be gathered in the field. *In-vitro* storage involves the establishment of cultures, preferably of a type that carry a minimal risk of somaclonal variation, and their transfer to slow growth or cryopreservation.

Slow growth is usually achieved by reducing the culture temperature or adding growth retardants to the culture medium. *In-vitro* base storage can only be achieved by cryopreservation, that is, storage in liquid nitrogen. All types of culture have been successfully crypreserved but cell cultures are the most easy to handle and shoot-tips the most problematic.

A new approach to cryopreservation termed 'vitrification' offers the potential" to overcome problems experienced with highly organized material such as shoot-tips. Another exciting new departure is the involvement of artificial seed technology in conservation work. The stage of development of *in-vitro* conservation varies from crop to crop with cassava being the most highly developed of the problem crops for which this approach is beneficial, and coconut being among the least well developed.

In these as in other cases, *in-vitro* conservation will never replace conventional technologies entirely but will complement them within a strategy that balances of the traditional approaches of *in-situ*, seed and field genebank conservation with the best of the new using where appropriate, *in-vitro* conservation, pollen storage and even DNA storage.

Chapter 8

Plant Genetic Resources Activities

The field of plant genetic resources (PGR) has gained much momentum, particularly in the last two decades. It is well realised now that no single country or region can be self-sufficient in its needs on genetic diversity. It is estimated that average inter-dependence between all regions of the world is more than 50 per cent - for some regions it may even be more than 90 per cent. By and large, mankind depends on barely 25 crops for its major needs.

The monoculture practices, environmental degradation and urban development - all have contributed to loss of plant genetic diversity, so vital to man's food security. The need to protect and conserve these resources more systematically is getting recognised now more and more, and over 100 national programmes are presently operating. However, with the spread of awareness in plant genetic resources in global context, several issues have emerged and plant genetic resources have become the subject of political and economic controversy.

Thus, global collaboration in PGR activities and their management is of vital importance The term 'Genetic Resources' started with the technical conference organised by the Food and Agriculture Organisation (FAO), wherein scientific principles, methodologies and strategies of exploration, conservation, evaluation and documentation were elaborately for the first time.

A follow up of this resulted in a report on the survey of crop genetic resources and through subsequent conferences, in two more very timely publications. Since then the activities have developed fast in national and international perspective. Here, it is intended to deal with the role of international organisations which have gone a long way in initiating, promoting and accelerating concern on plant genetic resources in all possible dimensions of PGR activities.

CONSULTATIVE GROUP ON INTERNATIONAL AGRICULTURAL RESEARCH (CGIAR)

The Consultative Group on International Agricultural Research

(CGIAR) was established in 1971. It is an association of countries, international and regional organisations and private foundations dedicated to supporting a system of agricultural research centres and programmes around the world.

The research carried out is directed to improve the quantity and quality of food production in developing countries. The World Bank, FAO and UNDP are co-sponsors of this effort. The CGIAR is advised by a Technical Advisory Committee (TAC).

Currently, 13 International Agricultural Research Centres (IARCs) are working under the aegis of the CGIAR. Of these, nine are crop commodity institutes and are also involved in the conservation and utilisation of plant genetic resources. However, the main Centre for PGR activities under the CGIAR is the International Board for Plant Genetic Resources (IBPGR).

The purpose of the CGIAR support to plant genetic resources is to ensure that the diversity of germplasm is safely maintained and made freely available for use in research and crop improvement for long-term benefit of all people. The CGIAR seeks to achieve this goal both directly through the institutions it supports and indirectly through strengthening national capabilities.

The PGR activities supported by the CGIAR are thus diverse and include exploration and collection, characterization, multiplication, evaluation, conservation, data management, information service and the supply of germplasm to plant breeders/research workers. Where appropriate, these activities are supported by research and training.

INTERNATIONAL AGRICULTURAL RESEARCH CENTRES (IARCS)

International Agricultural Research Centres (IARCs) of the CGIAR, the countries where these are situated and their mandate.Of these, nine are crop commodity centres. With well defined goals, the efforts of these IARCs directed towards augmenting genetic diversity in their mandate crops have been appreciable lists the germplasm accessions assembled by these IARCs which represent a sizable part of the total world holdings for the respective crops.

The world germplasm collections of food crops are based on recent analysis. This also indicates the percentage of genetic material in terms of landraces and wild species, and the possible degree of threat posed to material still to be collected. The degree of comprehensiveness of germplasm collections among major food crops may thus be ranked in the order of cereals, common or Irish potato, grain legumes, sorghum and millets, cassava, soybean and peanuts, sweet potato and yam.

It is also estimated that the duplication of accessions within a crop among the various genebanks may vary from 35%(in peanut) to 75 % (in wheat).Most of the IARCs now have or are in a process of getting established long-term conservation facilities. These centres are also equally concerned with the development of overall data bases on world holdings of their mandate crops.

The Importance and Relative Advantage of the IARCs are

- The maintenance of global collections of the mandate crops, their thorough evaluation and active utilisation in the breeding programmes, are all integrated programmes in one and the same institute.
- The existence of well managed cooperating networks with the national programmes mainly on crop improvement, but also more and more on PGR. This has facilitated the flow of germplasm and helped building-up capabilities/facilities in the participating countries.

The IBPGR has substantially assisted the IARCs to collect germplasm which is held in trust by the IARCs and in other PGR research activities. Since 1988, IBPGR and the concerned Genetic Resources Units of the IARCs meet regularly and PGR issues of common interest in an Inter-Centre PGR Working Group.

IBPGR and its Activities

The International Board for Plant Genetic Resources (IBPGR) was established in 1974 within the CGIAR system to promote and coordinate the collection, conservation, documentation, evaluation and use of germplasm.

Since its inception, IBPGR has been instrumental in accelerating germplasm collecting activities based on crop as well as regional priorities according to the loss of genetic diversity through spread of modern cultivars/HYVs or other bio-ecological causes resulting in the depletion of genetic diversity.

In 1986, the CGIAR approved an extended mandate to enable IBPGR to catalyze whatever actions were needed to support and widen the already existing global genetic resources network. This mandate is to further the study, collection, preservation, documentation, evaluation and utilisation of the genetic diversity of useful plants for the benefit of people throughout the world.

IBPGR shall act as a catalyst both within and outside the CGIAR system in stimulating the action needed to sustain a viable network of institutions for the conservation of genetic resources for these plants.

In order to meet the requirements as described in the mandate, IBPGR reconstructed its programme as follows:

Germplasm Acquisition

monitoring degree of genetic erosion, collecting threatened germplasm, supplementing existing gaps in germ-plasm collections and facilitating germplasm flow on a global scale.

Germplasm Characterization and Evaluation

Ggenerating the badly needed data on the existing collections through standardization of procedures to process, store and distribute characterization and evaluation data. Data acquisition, data analysis, application and evaluation strategy are the programme elements.

Training

The development of conceptual, technical and managerial skills through support of manpower training. This involves postgraduate training, specialized short technical courses, individual training and intern fellowships.

In-vitro Culture Research

Development of *in-vitro* techniques for the collection, conservation and exchange of genotypes for recalcitrant species including collecting and tissue culture technology, disease indexing and therapy, cryopreservation, genetic stability and a pilot study for *in-vitro* genebanks.

Genetic diversity research: to enhance our understanding on the origin, evolution and variation patterns of crop genepools. This includes species mapping, ecogeographic studies, development of biochemical methods of description and research on wild relatives in priority crop genepools.

Seed Conservation Research

To establish and implement standards for seed storage. This effort includes study of physiology of stored orthodox and recalcitrant seeds, their genetic stability, dormancy, regeneration and genetic integrity as well as non-destructive disease indexing.

Global Genetic Resources Network

Developmental activities with the cooperating national programmes and IARCs including fostering base and active collections and data management and transfer. The important role of crop germplasm networks is more and more realised and used as an organisational instrument to bring all concerned workers together.

The regional offices provide vital links within their area of operation between the IBPGR, national programmes and regional programmes in gearing PGR activities in the national plant genetic resources systems. They also effectively maintain coordination and cooperation with the IARCs and other regional/international organisations within their jurisdiction.

They organise periodic meetings of the national PGR coordinators of different countries within the region. Thus, these offices are playing greater coordinating role, advisory service and on-spot guidance to national programmes as per the needs of the latter. The professional staff at each office, Coordinator, Associate Coordinator or Assistant Coordinator, maintain effective dialogue and cooperation and are IBPGRs true life-lines in catalyzing and strengthening such activities.

Some of the major past achievements of the IBPGR are summarized below:

- The development of a global network of genebanks to conserve germplasm. At present, the network comprises over 100 operating genebanks in developing and developed countries.
- The stimulation of other institutions on national, regional and international levels to assume the responsibility to collect, characterize and store plant genetic resources. These genebanks now operate in over 100 countries and range from emergent national entities to highly sophisticated centres in the CGIAR and in some national programmes.
- The establishment of priorities by species and regions for the collection of threatened germplasm. Over 400 collection missions worldwide were organised and/or supported.
- The support to train over 1450 individuals in all aspects of genetic resources conservation and utilisation.
- The standardisation of characterization and evaluation activities through the production of descriptor lists for all major crop species (over 60 crop descriptors have been prepared).
- The publication and dissemination of scientific reports, directories and newsletters.

Global Network of Base Collections

Since 1976, the IBPGR has identified a number of base collection genebanks around the world to maintain collections of specific crops on a regional or global basis.

These base collections cover mainly cereals (wheat, rice, maize, barley, sorghum and millets, oats and rye); food legumes (chickpea, faba bean, groundnut, lentil, lupin, *Phaseolus,* pigeonpea, soybean, mung bean, cowpea and winged bean); rootcrops (cassava, potato and sweet potato); vegetables (*Allium,* amaranth, *Capsicum,* cruciferous crops, cucurbits, okra, tomato and egg plant); industrial crops (cotton, sugarcane and tobacco); and forage legumes and grasses (16 genera).

These genebanks hold the collections of specific crops under conditions which preserve their long-term viability. However, some of these base collections are held by centres with only medium-term storage conditions which are accordingly being encouraged to upgrade their facilities to attain the standards necessary for long-term seed storage.

Crop Genetic Resources Networks

The primary objective of a crop network is to improve and/or coordinate conservation and utilisation of the crop genepool starting with a dialogue between all concerned parties-germplasm collectors, curators, researchers, breeders and other users -following an integrated approach. Through stimulating the developments of such networks, IBPGR is aiming at:

- Long-term secure storage in base collections;
- Ready provision of information through databases; and
- Ready availability of samples from active collections linked with base collections and improved use of crop genepool.

The IBPGR will act as a catalyst in bringing the specialists together who then develop plans for the network. The coordination of such networks can be with an international centre or a lead breeding institute. In order to gain experience and develop the concept, a task force was established and a pilot programme was launched to promote this concern on eight crops, viz. barley, maize, groundnut, sweet potato, medics, banana, okra and sugarbeet.

Each network is being developed through meetings/workshops designed to bring together national and international bodies, base and active collection curators, researchers as well as the user community. Once these initially identified networks have been set up, IBPGR plans to devote more resources to creating others. Eventually, it is hoped that networks will be established for all major crops as well as for many of the minor species, and that these networks will become self-sustainable.

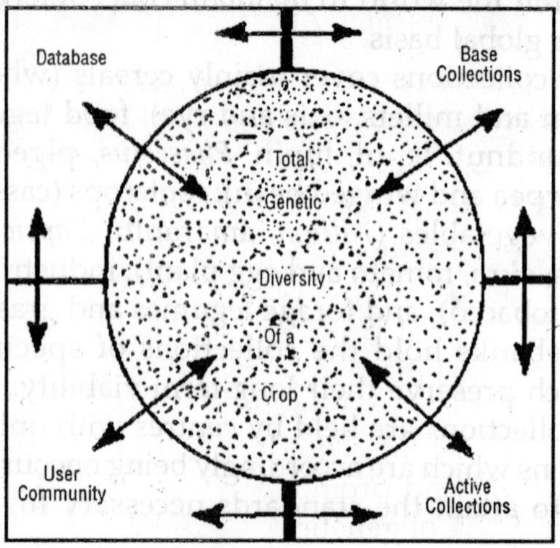

Fig. Schematic Plan of IBPGR's Concept of Crop Networks

Integrated conservation Strategies

The starting point for devising a conservation strategy is the composition of the genepool. It will be more likely that a balanced application of technologies will be needed in both *in-situ* and *ex-situ* conservation. Within the latter category, a further balance needs to be made between seed, field genebank, *in-vitro,* pollen and, perhaps in the future, DNA and gene storage.

Depending on the biological factors available (technical) infrastructure, amount of accessions at hand and some other factors, one will choose one or a combination of conservation methods for a given genepool. Examples of three selected crops demonstrating the differences between orthodox and recalcitrant seed crops between sexually and asexually propagated crops, between annual and perennial species, and the existence and/or distribution of the wild species in their natural habitat. In the latter case, the relevance by *in-situ* conservation is there, whereas a crop like coconut without related wild species does not require extensive *in-situ* conservation.

The methods, are not all suitable for long-term conservation. The majority are only suitable for short or medium-term conservation, and some others such as cryopreservation and DNA or gene libraries need further research to be safely utilised for long-term conservation. At present, only seed storage and, although less used so far, pollen storage, are safe and suitable methods enabling long-term static conservation of genetic diversity. In the case of *in-situ* conservation, one allows the naturally occurring species to evolve in their ecosystem and, if proper monitoring is available, this will be an adequate form of dynamic conservation.

PGR DATA MANAGEMENT

The IBPGR coordinates and promotes a large network of crop genetic resources activities and has to collect, organise and provide information concerning different activities within the network. This category of information covers manpower, infrastructure and organisation of national programmes, active projects, scientific literature, training, germplasm collection with IBPGR support, etc. The IBPGR documentation programme is thus multidimensional and currently consists of the following major components:

- Development and promotion of the use of standards for the documentation of genetic resources.
- Support to national and international programmes to establish and/or strengthen systems for efficient handling of the data, and to acquire adequate data on accessions in collections.
- Analysis and dissemination of genetic resources information on crops and other activities e.g. collecting, conservation, training, etc.

Central Crop Databases[SH]

Numerous advantages of integrating the data on germplasm held in different centres into one computer database were recognised by the IBPGR years ago.

The technological advances in the field of data processing has made implementation of such databases technically feasible at a reasonable cost.

Thus, in 1982, the IBPGR entered this new area of documentation of genetic resources, and the following policy was adopted with regard to establishment of centralised crop databases:

- The data bases should be organised separately for each crop.
- Crop databases should be located in an internationally recognised 'Centre of excellence' for the crops concerned.
- Databases for mandate crops of the IARCs of the CGIAR should preferably be dealt by the IARCs concerned.

A central database should enable both managers of genebanks and agencies coordinating genetic resource activities to target collecting efforts; see whether accessions are safely deposited in base collections and available from active collections; plan collaborative trials aiming at characterization and/or evaluation of germplasm etc. Thus, a database is a service to breeders and other users.

FAO AND PLANT GENETIC RESOURCES ACTIVITIES

The initial role of the FAO, much earlier to the establishment of the IBPGR and the activities of the IARCs, has already been pointed out. Since 1983, FAO has developed a global system in plant genetic resources.

- A flexible legal framework - the International Undertaking on Plant Genetic Resources.
- An intergovernmental forum - the Commission on Plant Genetic Resources.
- The beginning of a financial mechanism - the International Fund for Plant Genetic Resources.

The *International Undertaking on Plant Genetic Resources* is a formal arrangement, the objective of which is to ensure that plant genetic resources, especially species of present or future economic and social importance, will be explored, collected, preserved, evaluated and made available without restriction for breeding and other scientific purposes.

The *Commission on Plant Genetic Resources* is a unique intergovernmental global forum, where countries which are donors or users of germplasm as well as those which are sources of germplasm can plant genetic resources on an equal footing, and monitor the implementation of the principles contained in the International Undertaking.

The *International Fund For Plant Genetic Resources* is to help ensure the conservation and promote the use of plant genetic resources on a sustainable basis at world level. The Fund provides the mechanism for countries, intergovernmental and non-governmental organisations, and private industries and individuals to fulfil the common responsibility to maintain the world's plant genetic diversity.

To date, 122 countries have either joined the Commission (102), have agreed to adhere to the International Undertaking (90) or taken both steps. In short, the above aspects safeguard the conservation and use of biological diversity, in plant genes, genotypes and genepools; at molecular, population, species and ecosystem level.

FAO Commission on Plant Genetic Resources

The role of the International Commission on Plant Genetic Resources is to serve as a unique intergovernmental body that monitors the implementation of the Undertaking, and ensures the comprehensiveness and efficiency of the global system in full coordination with other national, regional and international organisations involved.

Since the establishment of the Commission in November 1983, and during the debates in its first two meetings which were attended both by member and non-member countries of the Commission the major reservations raised by some countries to the FAO Undertaking and Commission have been in regard to:

- The compatibility of the Undertaking with existing national laws related to Plant Breeders' Rights for many developed countries and to restriction of exchange of certain species in some developing countries.
- The possible overlap between the Commission and other organisations dealing with plant genetic resources. Apart from Plant Breeders' rights, Farmers' rights have also been recognised.

The International Undertaking recognises the rights of both plant breeders and farmers to benefit from their contributions to the improvement of agriculture. The adoption of Plant Breeders' rights has encouraged the development, marketing and widespread use of improved crop varieties but it has been partly responsible for reduction in the pool of crop germplasm conserved by growers.

The effects of the adoption of Farmers' rights are not yet evident but should result in improved funding for and appreciation of plant genetic resources. The International Commission on PGR will have the responsibility to provide a continuing scientific and technical contribution and debates on these topics and ensure that information of the highest standard is readily available as a basis for the formulation of national and international policies on germplasm availability and exchange.

International Agreements/Code of Conduct

The Commission considered the development of international agreements for the conservation and use of plant genetic resources.

In this context, it recommended that the Secretariat, in cooperation with the Working Group, should draft a code of conduct for international collectors of germplasm as also cover the conservation and use of PGR; and develop a code of conduct for biotechnology as it affects conservation and use of plant genetic resources.

FAO/IBPGR/IARCs Collaboration

The IARCs and IBPGR recognise the importance of collaboration and consultation with FAO on all aspects of plant genetic resources activities. It is further recognised that FAO and the FAO Commission are probably in a better position than IBPGR and IARCs to address a number of issues of a political nature, many of which require governmental legislation. Some specific areas for collaboration are:

- The Commission is to continue to follow-up on the free availability of germplasm and the support of all IARCs to the joint FAO/ IBPGR effort on the development of guidelines for the safe transfer of germplasm of specific crops.
- The establishment of an organisational framework for plant genetic resources work at the national and, if required, at the regional level. The IBPGR and IARCs have for some time recognised the need for national committees, programmes and coordinators, through which and with whom they can work. It is believed that it should be the responsibility of the Commission to encourage the countries to establish this structure.
- It is recognised that in a number of countries the development of national programmes requires technical assistance for institution building. The IBPGR and the IARCs can provide the scientific inputs in joining FAO and the Commission in mobilizing these funds.
- In the setting-up of crop networks, the Commission should be involved in obtaining the agreements for the designations of Centres to hold base collections. The technical aspects of coordinating the networks and filling the gaps in collecting and characterization can be undertaken by IBPGR/IARCs and the collaborating institutes.
- Safety duplication of base collections in permafrost could be another area for collaboration between the Commission, the IARCs and IBPGR.
- The designation of *in-situ* reserves is a national responsibility; the Commission should follow this up with the Governments concerned. IBPGR and IARCs can advise on the technical aspects of the identification and management of these areas.
- National capability in germplasm evaluation and plant breeding in developing countries requires strengthening and this is seen as a major task of the Commission and the FAO.

Other International/Regional Organisations

Besides the CGIAR network and the FAO, there are several other centres involved in PGR activities. The Asian Vegetable Research and Development Centre (AVRDC, Taiwan) has performed parallel activities as those of the IARCs.

The International Development Research Centre (IDRC) is also actively engaged in PGR utilisation and conservation - bamboos and rattans, banana, oilseeds, smaller millets; the International Jute Organisation (IJO) in promoting collection, utilisation and conservation of jute and kenaf genetic resources; the Japanese International Cooperation Agency (JICA); the German Agency for Technical Cooperation (GTZ); USAID and other assistance agencies in creating conservation facilities.

There is an International Network for the Improvement of Banana and Plantain (INIBAP, France), the Commonwealth Scientific and Industrial Research Organisation (CSIRO, Australia); well established country centres, such as the National Plant Germplasm System, USDA (USA), and the N.I. Vavilov All-Union Scientific Research Institute of Plant Industry/VIR (USSR); in Africa, the Plant Genetic Resources Centre/Ethiopia (PGRC/E); in Latin America, CENARGEN, Embrapa (Brazil); in East Asia, the Institute of Crop Germplasm Resources under the Chinese Academy of Agricultural Sciences (CAAS), Beijing; in South-East Asia, the National Plant Genetic Resources Laboratory, University of the Philippines at Los Baños, Philippines; and in South Asia, the National Bureau of Plant Genetic Resources (NBPGR), New Delhi, India.

Lately, the Commonwealth Science Council (CSC), UK, has also exhibited much concern and interest in this discipline, confined to lesser known plants/traditional useful plants-plants of ethnobotanical interest. The objectives of these centres/organisations may vary broadly depending on national/regional/international mandate but, by and large, the PGR activities carried out by them are complementary, with emphasis on conservation and utilisation of plant genetic wealth.

More recently, much concern on these activities has been highlighted, both globally and nationally, by several NGOs (Non-Governmental Organisations)/voluntary organisations such as RAFI, KENGRA, GRAIN, and in India, the Chipko movement.

BIODIVERSITY

Biological diversity is basic to both ecological and food security. Sustainable advances in biological productivity will not be possible without access to biological diversity (Swaminathan, 1989). Thus, safeguarding this endowment of nature is of paramount importance to human welfare. Its conservation is now receiving more and more attention.

Several United Nations agencies have become involved in this field including the FAO, United Nations Environment Programme (UNEP), and the United Nations Educational, Scientific and Cultural

Organisation (UNESCO). Many non-governmental organisations are involved on the conservation of biodiversity as well, both at the global and the national level. At the global level, this includes organisations, such as the World Wide Fund for Nature (WWF), the International Union for Conservation of Nature and Natural Resources (IUCN), the World Resources Institute (WRI), and many others.

Most of the organisations mentioned here are concerned about species loss rather than the loss of diversity within species. Nevertheless, there is a movement by these organisations to look below the species level as well, as shown in the 'Botanic Gardens Strategy'.

Botanic Gardens and Plant Genetic Resources Conservation

In a recent document prepared by the WWF and IUCN-BGCS, it has been pointed out that as many as 60,000 plant species may be in danger of extinction or serious genetic erosion during the next 30 to 40 years. Thus, multi-pronged efforts are required to halt such species extinction and promote their conservation, study and utilisation.

The botanical gardens have to play a pivotal role in this context with the triple purpose of conservation, propagation and public awareness. Their role in the last few centuries as centres of plant collection, introduction and distribution of material as well as seats of active research in systematic and economic botany has been well realised all over the world.

There are about 1500 botanic gardens and arboreta in the world, and these are playing their role in achieving conservation of plant life - wild species, cultivated types and as seeds in genebanks. The three main objectives of the living resources conservation are to:

- Maintain essential ecological processes and life support systems.
- Preserve genetic diversity.
- ensure that the utilisation of species and ecosystem is sustainable.

Currently, the IUCN Botanic Gardens Conservation Secretariat (BGCS) is operating. The Botanic Gardens Conservation Strategy provides a rationale for the involvement of botanic gardens in conservation, and gives policy guidance on how this can be achieved and the urgency of this task. It recommends that:

- Each individual garden clarifies its commitment to conservation in a Mission Statement and adopts more professional standards of management to achieve its Mission on conservation, documentation and exchange of information on plant species held.

- Provides the basis for a more coherent Accessions Policy that takes account of conservation needs and what plants are held in other botanic gardens.
- For *in-situ* conservation, it outlines the role of the garden in habitat evaluation, rare species monitoring, 'habitat gardening' and managing protected areas.
- For managed *ex-situ* conservation, proposes strict rules and procedures for the establishment of seed banks, field genebanks and other germplasm collections, and outlines methods of sampling populations to maintain adequate genetic variation.

It proposes that maximum emphasis be put on collaboration between gardens at a national level or, where appropriate, regional level. It recommends collaboration at international level through the IUCN Botanic Gardens Conservation Secretariat.

DEVELOPMENTS IN BIOTECHNOLOGY

The impact of biotechnology affects the accessibility of germplasm. There has been an active interest in the use of biotechnology for germplasm work for several years. *In-vitro* collecting of germplasm and its conservation under slow growth conditions are proving to be useful additional technologies for genetic resources work. Cryopreservation techniques are also under active investigation.

Analyses of species relationships in genepools using molecular genetic techniques have provided useful results to guide collecting and utilisation programmes. The development of reliable and sensitive disease indexing procedures using antibody techniques is helping the safe transfer of germplasm throughout the world. Thus, it is necessary to ensure that new techniques are adapted for wider application in plant genetic resources work. A pragmatic approach is required alongside more conventional technologies.

GENERAL CONSIDERATIONS

The international perspective on PGR activities, as these are taking shape today, is highlighted above. The linkages and mandates of different organisations have been reflected. It is quite evident that the problems relating to management, use and conservation of plant genetic resources are of diverse nature - scientific, technical, legal and political.

Safeguarding of crop germplasm assumes high priority. Much collections have been made and stored in genebanks. But inside agricultural experimental stations and genebanks genetic erosion continues at an alarming rate. This is particularly so in tropical countries where high temperature and humidity shorten the period during which seeds remain viable and where a considerable number of accessions can only be maintained in rather unsafe field genebanks at a high cost.

Preservation is often handicapped by lack of adequate cold storage facilities, meagre support for genebank operations, shortage of trained people or discontinuity of their service and poor financial or other incentives. These situations, which hamper genetic conservation, need to be improved.

There are, in IBPGR's network, around 140 base collections of crops and 50 centres/genebanks. This network now has spread to China, and links have been developed with the USSR. However, much needs to be done by major genebanks in comparing holdings, culling redundancy and filling in gaps. The goal should be to promote judicious exchange and consolidation among collections so that duplicate storage can be provided for every valid accession.

The quality of seeds entering storage, the efficacy of storage facilities, the extent of scientific activities, the level of management, and the security of conserved stocks of seeds vary considerably from one genebank to another. It is not surprising, therefore, that genetic erosion occurs in every genebank to some extent. A collection may suffer great losses, if it is improperly monitored for viability or if rejuvenation to produce fresh seeds is delayed.

Though, during the last two decades, the IARCs and the IBPGR have facilitated international exchange of germplasm, the scenario has changed with private seed firms/multinationals coming into picture and, in some cases, monopolising specific germplasm resources to their advantage. This has led to some apprehensions that breeding of elite cultivars might fall largely into a private sector which also produce agricultural chemicals.

Further, widespread government regulations related to the Plant Breeders' Rights could restrict the farmers from use of patented seeds and plants. This has been the subject of debate for quite sometime and recently, the Seed Association of India had organised an international symposium.

The IBPGR concern was highlighted Seed Association of India. The FAO Commission is also much involved as indicated above in these issues related to Plant Breeders' Rights and Farmers' Rights.

To mitigate the apprehensions of the developing countries, some of the activists/NGOs have widely publicised these issues which are still being debated. Possibly, steps such as the implementation of

International Undertaking on Plant Genetic Resources will provide a bridge to ensure free access to crop germplasm on a global basis.

Thus, much public awareness has been generated. The society is well aware now about the importance of germplasm and its use. Practical management and policy measures need to be worked out. The outcome of the second meeting in the Dialogue Series on PGR, held in Madras in early 90s.

The above issues apart, with all our achievements on this front, there is still *a* need to augment genetic diversity from protected and natural habitats, such as unimproved germplasm grown in traditional farmlands, orchards and forestlands. The national programmes are playing a more concerted role in this direction than ever before though these are frequently handicapped by funding aspects. These gene-rich countries would need an international umbrella to support their activities.

The IBPGR has been providing the required impetus in this direction through training programmes, distribution of information, guideline manuals, free advisory service and need-based support since its inception in collecting, conservation and documentation activities; though lately, it has taken up more of a catalyzing and advisory role. Certainly, national programmes, which are the backbone to the future build up and survival of this networking and coordination, must be strengthened further.

Concern on plant genetic resources has attained global, regional and national importance, particularly in the last two decades. The International Agricultural Research Centres (IARCs) are playing great role in the improvement and utilisation of their mandate crops, and in the collection and conservation of the concerned crop genetic diversity. However, the overall key role in promoting concern on plant genetic resources activities in a global context and in coordinating these activities is being performed by the International Board for Plant Genetic Resources (IBPGR) in close interaction with the FAO and IARCs, and the national programmes.

The FAO Commission on Plant Genetic Resources, as an inter-governmental body, is concerned with several policy issues to ensure the efficiency of operating a global system in full coordination with national, regional and international organisations. The role of other international organisations, such as the International Union for Conservation of Nature and Natural Resources (IUCN), the World Wide Fund for Nature (WWF) and several others has also been stressed.

Chapter 9

Wastewater Treatment Technologies

Relatively simple wastewater treatment technologies can be designed to provide low cost sanitation and environmental protection while providing additional benefits from the reuse of water. These technologies use natural aquatic and terrestrial systems.

Mechanical treatment systems, which use natural processes within a constructed environment, tend to be used when suitable lands are unavailable for the implementation of natural system technologies. Aquatic systems are represented by lagoons; facultative, aerated, and hydrograph controlled release (HCR) lagoons are variations of this technology. Further, the lagoon-based treatment systems can be supplemented by additional pre- or post-treatments using constructed wetlands, aquacultural production systems, and/or sand filtration.

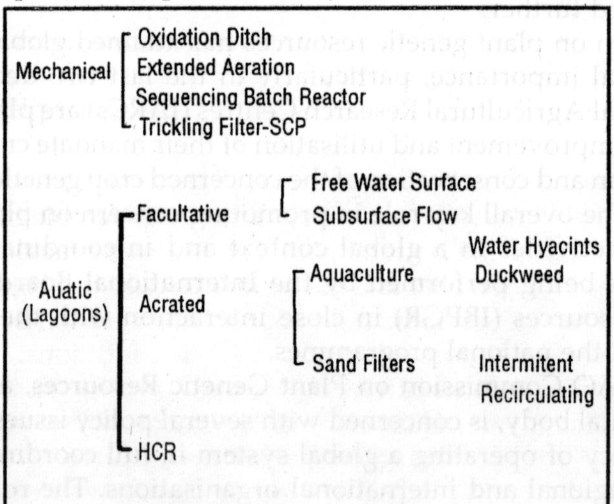

Fig. Summary of Wastewater Treatment Technologies.

They are used to treat a variety of wastewaters and function under a wide range of weather conditions. Terrestrial systems make use of the nutrients contained in wastewaters; plant growth and soil adsorption convert biologically available nutrients into less-available forms of biomass, which is then harvested for a variety of uses, including methane gas production, alcohol production, or cattle feed supplements.

TECHNICAL DESCRIPTION

Mechanical Treatment Technologies

Mechanical systems utilize a combination of physical, biological, and chemical processes to achieve the treatment objectives. Using essentially natural processes within an artificial environment, mechanical treatment technologies use a series of tanks, along with pumps, blowers, screens, grinders, and other mechanical components, to treat wastewaters.

Flow of wastewater in the system is controlled by various types of instrumentation. Sequencing batch reactors (SBR), oxidation ditches, and extended aeration systems are all variations of the activated-sludge process, which is a suspended-growth system. The trickling filter solids contact process (TF-SCP), in contrast, is an attached-growth system. These treatment systems are effective where land is at a premium.

Aquatic Treatment Technologies

Facultative lagoons are the most common form of aquatic treatment-lagoon technology currently in use. The water layer near the surface is aerobic while the bottom layer, which includes sludge deposits, is anaerobic. The intermediate layer is aerobic near the top and anaerobic near the bottom, and constitutes the facultative zone. Aerated lagoons are smaller and deeper than facultative lagoons. These systems evolved from stabilization ponds when aeration devices were added to counteract odors arising from septic conditions.

The aeration devices can be mechanical or diffused air systems. The chief disadvantage of lagoons is high effluent solids content, which can exceed 100 mg/l. To counteract this, hydrograph controlled release (HCR) lagoons are a recent innovation. In this system, wastewater is discharged only during periods when the stream flow is adequate to prevent water quality degradation. When stream conditions prohibit discharge, wastewater is accumulated in a storage lagoon.

Constructed wetlands, aquacultural operations, and sand filters are generally the most successful methods of polishing the treated wastewater effluent from the lagoons. These systems have also been used with more traditional, engineered primary treatment technologies such as Imhoff tanks, septic tanks, and primary clarifiers. Their main advantage is to provide additional treatment beyond secondary treatment where required. In recent years, constructed wetlands have been utilized in two designs: systems using surface water flows and systems using subsurface flows.

Both systems utilize the roots of plants to provide substrate for the growth of attached bacteria which utilize the nutrients present in the effluents and for the transfer of oxygen. Bacteria do the bulk of the work in these systems, although there is some nitrogen uptake by the plants. The surface water system most closely approximates a natural wetland. Typically, these systems are long, narrow basins, with depths of less than 2 feet, that are planted with aquatic vegetation such as bulrush (*Scirpus* spp.) or cattails (*Typha* spp.).

The shallow groundwater systems use a gravel or sand medium, approximately eighteen inches deep, which provides a rooting medium for the aquatic plants and through which the wastewater flows.

Table Typical Design Features Aquatic Treatment Units

Technology	Treatment goal	Detention Time (days)	Depth (feet)	Organic Loading (lb/ac/day)
Oxidation pond	Secondary	10-40	3-4.5	36-110
Facultative pond	Secondary	25-180	4.5-7.5	20-60
Aerated pond	Secondary, polishing	7-20	6-18	45-180
Storage pond, HCR pond	Secondary, storage, polishing	100-200	9-15	20-60
Root zone Treatment, Hyacinth pond	Secondary	30-50	<4.5	<45

Aquaculture systems are distinguished by the type of plants grown in the wastewater holding basins. These plants are commonly water hyacinth (*Eichhornia crassipes*) or duckweed (*Lemna* spp.). These systems are basically shallow ponds covered with floating plants that detain wastewater at least one week.

The main purpose of the plants in these systems is to provide a suitable habitat for bacteria which remove the vast majority of dissolved nutrients. The design features of such systems are summarized in Table.

Table. Typical Design Features for Constructed Wetlands

Design Factor	Surface water flow	Subsurface water flow
Minimum surface area	23-115 ac/mgd	2.3-46 ac/mgd
Maximum water depth	Relatively shallow	Water level below ground surface
Bed depth	Not applicable	12.30m
Minimum hydraulic residence time	7 days	7 days
Maximum hydraulic loading rate	0.2-1.0 gpd/sq ft	0.5-10 gpd/sq ft
Minimum pretreatment	Primary (secondary optional)	Primary
Range of organic loading as BOD	9-18 lb/ac/d	1.8-140 lb/ac/d

Sand filters have been used for wastewater treatment purposes for at least a century in Latin America and the Caribbean. Two types of sand filters are commonly used: intermittent and recirculating. They differ mainly in the method of application of the wastewater. Intermittent filters are flooded with wastewater and then allowed to drain completely before the next application of wastewater.

In contrast, recirculating filters use a pump to recirculate the effluent to the filter in a ratio of 3 to 5 parts filter effluent to 1 part raw wastewater.

Both types of filters use a sand layer, 2 to 3 feet thick, underlain by a collection system of perforated or open joint pipes enclosed within graded gravel. Water is treated biologically by the epiphytic flora associated with the sand and gravel particles, although some physical filtration of suspended solids by the sand grains and some chemical adsorption onto the surface of the sand grains play a role in the treatment process.

Terrestrial Treatment Technologies

Terrestrial treatment systems include slow-rate overland flow, slow-rate subsurface infiltration, and rapid infiltration methods. In addition to wastewater treatment and low maintenance costs, these systems may yield additional benefits by providing water for groundwater recharge, reforestation, agriculture, and/or livestock pasturage. They depend upon physical, chemical, and biological reactions on and within the soil. Slow-rate overland flow systems require vegetation, both to take up nutrients and other contaminants and to slow the passage of the effluent across the land surface to ensure maximum contact times between the effluents and the plants/soils.

Slow-rate subsurface infiltration systems and rapid infiltration systems are "zero discharge" systems that rarely discharge effluents directly to streams or other surface waters. Each system has different constraints regarding soil permeability.

Although slow-rate overland flow systems are the most costly of the natural systems to implement, their advantage is their positive impact on sustainable development practices.

In addition to treating wastewater, they provide an economic return from the reuse of water and nutrients to produce marketable crops or other agriculture products and/or water and fodder for livestock.

The water may also be used to support reforestation projects in water-poor areas.

In slow-rate systems, either primary or secondary wastewater is applied at a controlled rate, either by sprinklers or by flooding of furrows, to a vegetated land surface of moderate to low permeability. The wastewater is treated as it passes through the soil by filtration, adsorption, ion exchange, precipitation, microbial action, and plant uptake.

Vegetation is a critical component of the process and serves to extract nutrients, reduce erosion, and maintain soil permeability.

Overland flow systems are a land application treatment method in which treated effluents are eventually discharged to surface water. The main benefits of these systems are their low maintenance and low technical manpower requirements.

Wastewater is applied intermittently across the tops of terraces constructed on soils of very low permeability and allowed to sheet-flow across the vegetated surface to the runoff collection channel. Treatment, including nitrogen removal, is achieved primarily through sedimentation, filtration, and biochemical activity as the wastewater flows across the vegetated surface of the terraced slope.

Loading rates and application cycles are designed to maintain active microorganism growth in the soil. The rate and length of application are controlled to minimize the occurrence of severe anaerobic conditions, and a rest period between applications is needed. The rest period should be long enough to prevent surface ponding, yet short enough to keep the microorganisms active. Site constraints relating to land application technologies are shown in Table.

Table. Site Constraints for Land Application Technologies

Feature	Slow Rate	Rapid Infiltration	Subsurface Infiltration	Overland Flow
Soil texture	Sandy loam to clay loam	Sand and sandy loam	Sand to clayey loam	Silty loam and clayey loam
Depth to groundwater	3 ft	3 ft	3 ft	Not critical
Vegetation	Required	Optional	Not applicable	Required
Climatic restrictions	Growing season	None	None	Growing season
Slope	<20%, cultivated land < 40%, uncultivated land	Not critical	Not applicable	2%-8% finished slopes

In rapid infiltration systems, most of the applied wastewater percolates through the soil, and the treated effluent drains naturally to surface waters or recharges the groundwater. Their cost and manpower requirements are low. Wastewater is applied to soils that are moderately or highly permeable by spreading in basins or by sprinkling. Vegetation is not necessary, but it does not cause a problem if present. The major treatment goal is to convert ammonia nitrogen in the water to nitrate nitrogen before discharging to the receiving water.

Subsurface infiltration systems are designed for municipalities of less than 2,500 people. They are usually designed for individual homes (septic tanks), but they can be designed for clusters of homes. Although they do require specific site conditions, they can be low-cost methods of wastewater disposal.

Extent of Use

These treatment technologies are widely used in Latin America and the Caribbean. Combinations of some of them with wastewater reuse technologies have been tested in several countries. Colombia has extensively tested aerobic and anaerobic mechanical treatment systems. Chile, Colombia, and Barbados have used activated sludge plants, while Brazil has utilized vertical reactor plants. Argentina, Bolivia, Colombia, Guatemala, Brazil, Chile, Curaçao, Mexico, Jamaica, and Saint Lucia have successfully experimented with different kinds of terrestrial and aquatic treatment systems for the treatment of wastewaters.

Curaçao, Mexico, and Jamaica have used stabilization or facultative lagoons and oxidation ponds; their experience has been that aquatic treatment technologies require extensive land areas and relatively long retention times, on the order of 7 to 10 days, to adequately treat wastewater.

An emerging technology, being tested in a number of different countries, is a hybrid aquatic-terrestrial treatment system that uses wastewaters for hydroponic cultivation. However, most of the applications of this hybrid technology to date have been limited to the experimental treatment of small volumes of wastewater.

Operation and Maintenance

Operation and maintenance requirements vary depending on the particular technology used. In mechanical activated-sludge plants, maintenance requirements consist of periodically activating the sludge pumps, inspecting the system to ensure that are no blockages or leakages in the system, and checking BOD and suspended solids concentrations in the plant effluent to ensure efficient operation.

In the case of aquatic treatment systems using anaerobic reactors and facultative lagoons for primary wastewater treatment, the following operational guidelines should be followed:

- Periodically clean the sand removal system (usually every 5 days in dry weather, and every 2 to 3 days in wet weather).
- Daily remove any oily material that accumulates in the anaerobic reactor.
- Daily remove accumulated algae in the facultative lagoons.
- Open the sludge valves to send the sludge to the drying beds.
- Establish an exotic aquatic plant removal program (aquatic plant growth can hamper the treatment capacity of the lagoons).
- Properly dispose of the materials removed, including dried sludge.

A preventive maintenance program should also be established to increase the efficiency of the treatment systems and prolong their lifespan.

When using terrestrial treatment systems or hybrid hydroponic cultivation systems for wastewater treatment, it is advisable to have two parallel systems, and to alternate applications of wastewater to these systems every 12 hours in order to facilitate aeration and to avoid damage to the system.

Care is required to avoid hydraulic overload in these systems, as the irrigated plant communities could be damaged and the degree of treatment provided negated. Periodic removal of sediments accumulated in the soil is also required to improve the soil-plant interaction and to avoid soil compaction/subsidence.

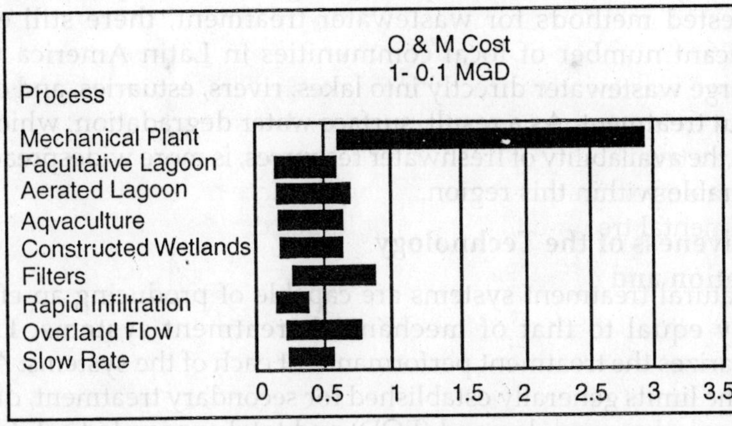

Fig. Comparative Operation and Maintenance Cost of Wastewater Treatment Technologies.

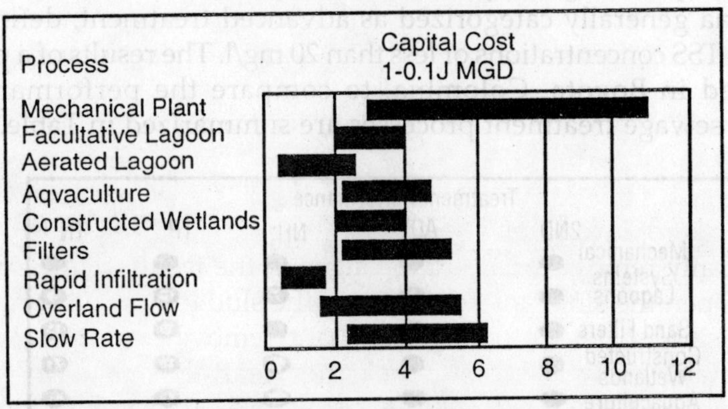

Fig. Comparative Capital Cost of Wastewater Treatment Technologies

LEVEL OF INVOLVEMENT

Government involvement is essential in the implementation of most of the wastewater treatment technologies. The private sector, particularly the tourism industry, has successfully installed "packaged" or small-scale, self-contained sewage treatment plants at individual sites. In some cases, the installation of these plants has been combined with the reuse of the effluent for watering golf courses, lawns, and similar areas.

The selection and construction of the appropriate wastewater treatment technology is generally initiated and financed, at least partially, by the government, with the subsequent operation and maintenance of the facility being a responsibility of the local community. Nevertheless, despite the large number of well-known and well-tested methods for wastewater treatment, there still exist a significant number of local communities in Latin America which discharge wastewater directly into lakes, rivers, estuaries, and oceans without treatment. As a result, surface water degradation, which also affects the availability of freshwater resources, is more widespread than is desirable within this region.

Effectiveness of the Technology

Natural treatment systems are capable of producing an effluent quality equal to that of mechanical treatment systems. **Fig.** 31 summarizes the treatment performance of each of the systems. All can meet the limits generally established for secondary treatment, defined as biological oxygen demand (BOD) and total suspended solids (TSS) concentrations of less than 30 mg/l.

All except the lagoon systems can also produce effluents that meet the criteria generally categorized as advanced treatment, defined as BOD and TSS concentrations of less than 20 mg/l. The results of a project conducted in Bogota, Colombia, to compare the performance of different sewage treatment processes are summarized in Table.

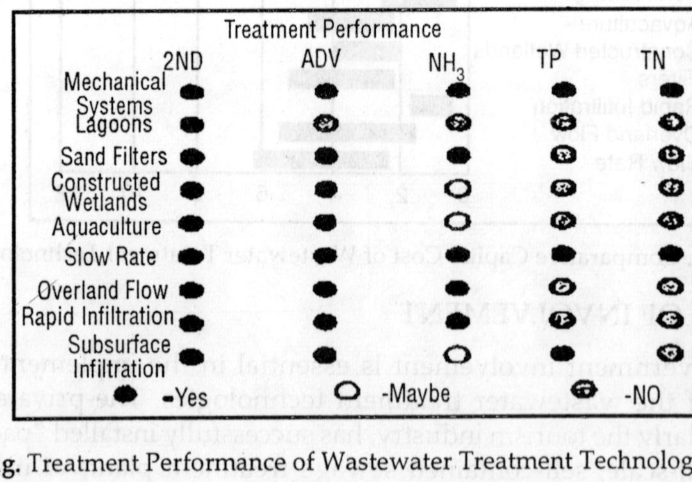

Fig. Treatment Performance of Wastewater Treatment Technologies.

Suitability

Mechanical systems are more suitable for places where land

availability is a concern, such as hotels and residential areas. Mechanical plants are the least land intensive of the wastewater treatment methods based on natural processes.

Lagoon and oxidation pond technologies are suitable where there is plenty of land available. Slow-rate systems require as much as 760 acres. Hybrid hydroponic cultivation techniques, using aquatic and terrestrial plants for the treatment for wastewater, also require relatively large amounts of land, and are best suited to regions where suitable aquatic plants can grow naturally.

Advantages

Table summarizes the advantages of the various wastewater treatment technologies. In general, the advantages of using natural biological processes relate to their "low-tech/no-tech" nature, which means that these systems are relatively easy to construct and operate, and to their low cost, which makes them attractive to communities with limited budgets.

Table. Comparative Performance of Sewage Treatment Systems

Process	Oxygen Supply	Reactor Volume	Retention Time	Removal Efficiency
Activated sludge	Pressurized air	10 m^3	4-6 hr	90%-95% organic matter 90%-95% suspend solids
Biologic rotary discs	Air	1 m^3	1-3 hr	90%-95% organic matter
Ascendant flow	Anaerobic	2 m^3	24 hr	50%-60% organic matter 57% suspended solids
Anaerobic filtration	Anaerobic	2 m^3	36 hr	40%-50% organic matter 52% suspended solids
Septic tank	Anaerobic	2 m^3	36 hr	25% organic matter
Hydroponic cultivation	Aerobic/anaerobic	6 m^3	12 hr	65%-75% organic matter

However, their simplicity and low cost may be deceptive in that the systems require frequent inspections and constant maintenance to ensure smooth operation. Concerns include hydraulic overloading, excessive plant growth, and loss of exotic plants to natural watercourses.

For this reason, and also because of the land requirements for biologically based technologies, many communities prefer mechanically-based technologies, which tend to require less land and permit better control of the operation. However, these systems generally have a high cost and require more skilled personnel to operate them.

Disadvantages

Table also summarizes the disadvantages of the various wastewater treatment technologies. These generally relate to the cost of construction and ease of operation. Mechanical systems can be costly to build and operate as they require specialized personnel. Nevertheless, they do offer a more controlled environment which produces a more consistent quality of effluent.

Natural biological systems, on the other hand, are more land-intensive, require less-skilled operators, and can produce effluents of variable quality depending on time of year, type of plants, and volume of wastewater loading. Generally, the complexity and cost of wastewater treatment technologies increase with the quality of the effluent produced.

Cultural Acceptability

Governments and the private sector in many Latin American countries fail to fully recognize the necessity of wastewater treatment and the importance of water quality in improving the quality of life of existing and future generations.

The contamination of natural resources is a major impediment to achieving the stated objective of Agenda 21 of environmentally sustainable economic growth and development.

Further Development of the Technology

The cost-effectiveness of all wastewater treatment technologies needs to be improved. New designs of mechanical systems which address this concern are being introduced by the treatment plant manufacturing industry. The use of vertical reactors with an activated-sludge system, being tested in Brazil in order to acquire data for future improvement of this technology, is one example of the innovation going on in the industry.

Similar product development is occurring in the use of aquatic and terrestrial plants and hybrid hydroponic systems, as a means of wastewater treatment; however, these technologies are still in an experimental phase and will require more testing and research prior to being accepted as standard treatment technologies. In addition, education to create an awareness of the need for wastewater treatment remains a critical need at all levels of government and

Table. Advantages and Disadvantages of Conventional and Non-conventional Wastewater Treatment Technologies

Treatment Type	Advantages	Disadvantages
Aquatic Systems		
Stabilization lagoons	Low capital cost Low operation and maintenance costs Low technical manpower requirement	Requires a large area of land May produce undesirable odors
Aerated lagoons	Requires relatively little land area Produces few undesirable odors	Requires mechanical devices to aerate the basins Produces effluents with a high suspended solids concentration

WASTEWATER REUSE

Once freshwater has been used for an economic or beneficial purpose, it is generally discarded as waste. In many countries, these wastewaters are discharged, either as untreated waste or as treated effluent, into natural watercourses, from which they are abstracted for further use after undergoing "self-purification" within the stream. Through this system of indirect reuse, wastewater may be reused up to a dozen times or more before being discharged to the sea.

Such indirect reuse is common in the larger river systems of Latin America. However, more direct reuse is also possible: the technology to reclaim wastewaters as potable or process waters is a technically feasible option for agricultural and some industrial purposes (such as for cooling water or sanitary flushing), and a largely experimental option for the supply of domestic water.

Wastewater reuse for drinking raises public health, and possibly religious, concerns among consumers. The adoption of wastewater treatment and subsequent reuse as a means of supplying freshwater is also determined by economic factors.

In many countries, water quality standards have been developed governing the discharge of wastewater into the environment.

Terrestrial Systems		
Septic tanks	Can be used by individual households Easy to operate and maintain Can be built in rural areas	Provides a low treatment efficiency Must be pumped occasionally Requires a landfill for periodic disposal of sludge and septage
Constructed wetlands	Removes up to 70% of solids and bacteria Minimal capital cost Low operation and maintenance requirements and costs	Remains largely experimental Requires periodic removal of excess plant material Best used in areas where suitable native plants are available
Mechanical Systems		
Filtration systems	Minimal land requirements; can be used for household-scale treatment Relatively low cost Easy to operate	Requires mechanical devices
Vertical biological reactors	Highly efficient treatment method Requires little land area Applicable to small communities for local-scale treatment and to big cities for regional-scale treatment	High cost Complex technology Requires technically skilled manpower for operation and maintenance Needs spare-parts-availability Has a high energy requirement
Activated sludge	Highly efficient treatment method Requires little land area Applicable to small communities for local-scale treatment and to big cities for regional-scale treatment	High cost Requires sludge disposal area (sludge is usually land-spread) Requires technically skilled manpower for operation and maintenance

Wastewater, in this context, includes sewage effluent, stormwater runoff, and industrial discharges. The necessity to protect the natural environment from wastewater-related pollution has led to much improved treatment techniques.

Extending these technologies to the treatment of wastewaters to potable standards was a logical extension of this protection and augmentation process.

Technical Description

One of the most critical steps in any reuse program is to protect the public health, especially that of workers and consumers. To this end, it is most important to neutralize or eliminate any infectious agents or pathogenic organisms that may be present in the wastewater.

For some reuse applications, such as irrigation of non-food crop plants, secondary treatment may be acceptable. For other applications, further disinfection, by such methods as chlorination or ozonation, may be necessary. Table presents a range of typical survival times for potential pathogens in water and other media.

Table. Typical Pathogen Survival Times at 20 - 30°C (in days)

Pathogen	Freshwater and sewage	Crops	Soil
Viruses	< 120 but usually <50	<60 but usually < 15	<100 but usually <20
Bacteria	<60 but usually <30	<3 0 but usually < 15	<70 but usually <20
Protozoa	<30 but usually <15	<10 but usually <2	<70 but usually <20
Helminths	Many months	<60 but usually <30	Many months

A typical example of wastewater reuse is the system at the Sam Lords Castle Hotel in Barbados. Effluent consisting of kitchen, laundry, and domestic sewage ("gray water") is collected in a sump, from which it is pumped, through a comminutor, to an aeration chamber. No primary sedimentation is provided in this system, although it is often desirable to do so.

The aerated mixed liquor flows out of the aeration chamber to a clarifier for gravity separation. The effluent from the clarifier is then passed through a 16-foot-deep chlorine disinfection chamber before it is pumped to an automatic sprinkler irrigation system. The irrigated areas are divided into sixteen zones; each zone has twelve sprinklers. Some areas are also provided with a drip irrigation system.

Sludge from the clarifier is pumped, without thickening, as a slurry to suckwells, where it is disposed of. Previously the sludge was pumped out and sent to the Bridgetown Sewage Treatment Plant for further treatment and additional desludging.

Extent of Use

For health and aesthetic reasons, reuse of treated sewage effluent is presently limited to non-potable applications such as irrigation of non-food crops and provision of industrial cooling water. There are no known direct reuse schemes using treated wastewater from sewerage systems for drinking. Indeed, the only known systems of this type are experimental in nature, although in some cases treated wastewater is reused indirectly, as a source of aquifer recharge.

Table presents some guidelines for the utilization of wastewater, indicating the type of treatment required, resultant water quality specifications, and appropriate setback distances. In general, wastewater reuse is a technology that has had limited use, primarily in small-scale projects in the region, owing to concerns about potential public health hazards.

Wastewater reuse in the Caribbean is primarily in the form of irrigation water. In Jamaica, some hotels have used wastewater treatment effluent for golf course irrigation, while the major industrial water users, the bauxite/alumina companies, engage in extensive recycling of their process waters.

In Barbados, effluent from an extended aeration sewage treatment plant is used for lawn irrigation. Similar use of wastewater occurs on Curaçao.

In Latin America, treated wastewater is used in small-scale agricultural projects and, particularly by hotels, for lawn irrigation. In Chile, up to 220 l/s of wastewater is used for irrigation purposes in the desert region of Antofagasta. In Brazil, wastewater has been extensively reused for agriculture. Treated wastewaters have also been used for human consumption after proper disinfection, for industrial processes as a source of cooling water, and for aquaculture.

Wastewater reuse for aquacultural and agricultural irrigation purposes is also practiced in Lima, Peru. In Argentina, natural systems are used for wastewater treatment. In such cases, there is an economic incentive for reusing wastewater for reforestation, agricultural, pasturage, and water conservation purposes, where sufficient land is available to do so.

Perhaps the most extensive reuse of wastewater occurs in Mexico, where there is large-scale use of raw sewage for the irrigation of parks and the creation of recreational lakes.

Table. Guidelines for Water Reuse

Type of Reuse	Treatment Required	Reclaimed Water Quality	Recommended Monitoring	Setback Distances
AGRICULTURAL	Secondary	pH = 6-9	pH weekly	300 ft from potable water supply wells
Food crops commercially processed	Disinfection	BOD ≤ 30 mg/l	BOD weekly	
		SS = 30 mg/l	SS daily	
Orchards and Vinerds		FC ≤ 200/100 ml	FC daily	100 ft from areas accessible to public
		Cl_2 residual = 1 mg/l min.	Cl_2 residual continuous	
PASTURAGE	Secondary	pH = 6-9	pH weekly	300 ft from potable water supply wells
Pasture for milking animals	Disinfection	BOD ≤ 30 mg/l	BOD weekly	
		SS ≤ 30 mg/l	SS daily	
Pasture for livestock		FC ≤ 200/100 ml	FC daily	100 ft from areas accessible to public
		Cl_2 residual = 1 mg/l min.	Cl_2 residual continuous	
FORESTATION	Secondary	pH = 6-9	pH weekly	300 ft from potable water supply wells
	Disinfection	BOD ≤ 30 mg/l	BOD weekly	
		SS ≤ 30 mg/l	SS daily	
		FC ≤ 200/100 ml	FC daily	100 ft from areas accessible to the public
		Cl_2 residual = 1 mg/l min.	Cl_2 residual continuous	
AGRICULTURAL	Secondary	pH = 6-9	pH weekly	50 ft from potable water supply wells
Food crops not commercially processed	Filtration	BOD ≤ 30 mg/l	BOD weekly	
	Disinfection	Turbidity ≤ 1 NTU	Turbidity daily	
		FC = 0/100 ml	FC daily	
		Cl_2 residual = 1 mg/l min.	Cl_2 residual continuous	
GROUNDWATER RECHARGE	Site-specific and use-dependent	Site-specific and use-dependent	Depends on treatment and use	Site-specific

In the United States, the use of reclaimed water for irrigation of food crops is prohibited in some states, while others allow it only if the crop is to be processed and not eaten raw. Some states may hold, for example, that if a food crop is irrigated in such a way that there is no contact between the edible portion and the reclaimed water, a disinfected, secondary-treated effluent is acceptable. For crops that are eaten raw and not commercially processed, wastewater reuse is more restricted and less economically attractive. Less stringent requirements are set for irrigation of non-food crops.

International water quality guidelines for wastewater reuse have been issued by the World Health Organization (WHO). Guidelines should also be established at national level and at the local/project level, taking into account the international guidelines. Some national standards that have been developed are more stringent than the WHO guidelines. In general, however, wastewater reuse regulations should be strict enough to permit irrigation use without undue health risks, but not so strict as to prevent its use. When using treated wastewater for irrigation, for example, regulations should be written so that attention is paid to the interaction between the effluent, the soil, and the topography of the receiving area, particularly if there are aquifers nearby.

Operation and Maintenance

The operation and maintenance required in the implementation of this technology is related to the previously discussed operation and maintenance of the wastewater treatment processes, and to the chlorination and disinfection technologies used to ensure that pathogenic organisms will not present a health hazard to humans.

Additional maintenance includes the periodic cleaning of the water distribution system conveying the effluent from the treatment plant to the area of reuse; periodic cleaning of pipes, pumps, and filters to avoid the deposition of solids that can reduce the distribution efficiency; and inspection of pipes to avoid clogging throughout the collection, treatment, and distribution system, which can be a potential problem.

Further, it must be emphasized that, in order for a water reuse program to be successful, stringent regulations, monitoring, and control of water quality must be exercised in order to protect both workers and the consumers.

Level of Involvement

The private sector, particularly the hotel industry and the

agricultural sector, are becoming involved in wastewater treatment and reuse. However, to ensure the public health and protect the environment, governments need to exercise oversight of projects in order to minimize the deleterious impacts of wastewater discharges.

One element of this oversight should include the sharing of information on the effectiveness of wastewater reuse. Government oversight also includes licensing and monitoring the performance of the wastewater treatment plants to ensure that the effluent does not create environmental or health problems.

Costs

Cost data for this technology are very limited. Most of the data relate to the cost of treating the wastewater prior to reuse. Additional costs are associated with the construction of a dual or parallel distribution system. In many cases, these costs can be recovered out of the savings derived from the reduced use of potable freshwater (i.e., from not having to treat raw water to potable standards when the intended use does not require such extensive treatment).

The feasibility of wastewater reuse ultimately depends on the cost of recycled or reclaimed water relative to alternative supplies of potable water, and on public acceptance of the reclaimed water. Costs of effluent treatment vary widely according to location and level of treatment. The degree of public acceptance also varies widely depending on water availability, religious and cultural beliefs, and previous experience with the reuse of wastewaters.

Effectiveness of the Technology

The effectiveness of the technology, while difficult to quantify, is seen in terms of the diminished demand for potable-quality freshwater and, in the Caribbean islands, in the diminished degree of degradation of water quality in the near-shore coastal marine environment, the area where untreated and unreclaimed wastewaters were previously disposed.

The analysis of beach waters in Jamaica indicates that the water quality is better near the hotels with wastewater reuse projects than in beach areas where reuse is not practiced: From an aesthetic point of view, also, the presence of lush vegetation in the areas where lawns and plants are irrigated with reclaimed wastewater is further evidence of the effectiveness of this technology.

Suitability

This technology has generally been applied to a small-scale

projects, primarily in areas where there is a shortage of water for supply purposes. However, this technology can be applied to larger-scale projects. In many developing countries, especially where there is a water deficit for several months of the year, implementation of wastewater recycling or reuse by industries can reduce demands for water of potable quality, and also reduce impacts on the environment.

Large-scale wastewater reuse can only be contemplated in areas where there are reticulated sewerage and/or stormwater systems. (Micro-scale wastewater reuse at the household or farmstead level is a traditional practice in many agricultural communities that use night soils and manures as fertilizers.) Urban areas generally have sewerage systems, and, while not all have stormwater systems, those that do are ideal localities for wastewater reuse schemes.

Wastewater for reuse must be adequately treated, biologically and chemically, to ensure the public health and environmental safety. The primary concerns associated with the use of sewage effluents in reuse schemes are the presence of pathogenic bacteria and viruses, parasite eggs, worms, and helminths (all biological concerns) and of nitrates, phosphates, salts, and toxic chemicals, including heavy metals (all chemical concerns) in the water destined for reuse.

Advantages

- This technology reduces the demands on potable sources of freshwater.
- It may reduce the need for large wastewater treatment systems, if significant portions of the waste stream are reused or recycled.
- The technology may diminish the volume of wastewater discharged, resulting in a beneficial impact on the aquatic environment.
- Capital costs are low to medium, for most systems, and are recoverable in a very short time; this excludes systems designed for direct reuse of sewage water.
- Operation and maintenance are relatively simple except in direct reuse systems, where more extensive technology and quality control are required.
- Provision of nutrient-rich wastewaters can increase agricultural production in water-poor areas.
- Pollution of seawater, rivers, and groundwaters may be reduced.
- Lawn maintenance and golf course irrigation is facilitated in resort areas.

- In most cases, the quality of the wastewater, as an irrigation water supply, is superior to that of well water.

Disadvantages

- If implemented on a large scale, revenues to water supply and wastewater utilities may fall as the demand for potable water for non-potable uses and the discharge of wastewaters is reduced.
- Reuse of wastewater may be seasonal in nature, resulting in the overloading of treatment and disposal facilities during the rainy season; if the wet season is of long duration and/or high intensity, the seasonal discharge of raw wastewaters may occur.
- Health problems, such as water-borne diseases and skin irritations, may occur in people coming into direct contact with reused wastewater.
- Gases, such as sulfuric acid, produced during the treatment process can result in chronic health problems.
- In some cases, reuse of wastewater is not economically feasible because of the requirement for an additional distribution system.
- Application of untreated wastewater as irrigation water or as injected recharge water may result in groundwater contamination.

Cultural Acceptability

A large percentage of domestic water users are afraid to use this technology to supply of potable water (direct reuse) because of the potential presence of pathogenic organisms. However, most people are willing to accept reused wastewater for golf course and lawn irrigation and for cooling purposes in industrial processes. On the household scale, reuse of wastewaters and manures as fertilizer is a traditional technology.

Further Development of the Technology

Expansion of this technology to large-scale applications should be encouraged. Cities and towns that now use mechanical treatment plants that are difficult to operate, expensive to maintain, and require a high skill level can replace these plants with the simpler systems; treated wastewater can be reused to irrigate crops, pastures, and lawns.

In new buildings, plumbing fixtures can be designed to reuse wastewater, as in the case of using gray water from washing machines and kitchen sinks to flush toilets and irrigate lawns. Improved public education to ensure awareness of the technology and its benefits, both environmental and economic, is recommended.

Microbial Fuel Cell technology for a Waste Water Treatment

Renewable energy is an increasing need in our society. Microbial fuel cell (MFC) technology represents a new form of renewable energy by generating electricity from what would otherwise be considered waste. This technology can use bacterium already present in wastewater as catalysts to generating electricity while simultaneously treating wastewater. Although MFCs generate a lower amount of energy than hydrogen fuel cells, a combination of both electricity production and wastewater treatment would reduce the cost of treating primary effluent wastewater.

Currently, most of the research performed on MFCs is concerned with increasing the power density of the system with respect to the peripheral anode surface area, while little research has been done on determining the size of a microbial fuel cell needed for a typical waste water treatment facility. The power density produced in a single chamber MFC can be modeled as a function of substrate concentration using an empirical Monod-type equation.

This equation exhibits power density as a state variable, the substrate concentration as the independent variable and maximum power density and the half-saturation constant as two parameters. This particular equation uses empirical data of the power density and substrate concentration needed to solve for the parameters of maximum power density and the half-saturation constant. These parameters are then used to model a function of power density, which is the state variable, versus substrate concentration, which is the independent variable.

2 PROBLEM FORMULATION

The objective of this study is to determine a size of the microbial fuel cell needed for the Arcata Waste Water Treatment Facility and its associated production costs. Analysis will include solving for the two parameters of maximum power density and the half-saturation constant within the Monod-type equation by use of the Newton-Raphson numerical method for hon-linear equations in a FORTRAN 90 computer program.

Assumptions pertaining to the MFCs characteristics and general design will be incorporated in the design. The information obtained from this feasibility study will be useful for integrating this alternative into future design proposals.

Biological Mechanism

In normal microbial catabolism, a substrate such as a carbohydrate is initially oxidized anaerobically when its electrons are released by enzymatic reactions.

The electrons are stored as intermediates (e.g., NADH, quinones) which become reduced and are then used to provide the living cell with energy. The ending location for the electrons is molecular oxygen or dioxygen at the end of the respiratory chain .

A MFC uses bacteria to catalyze the conversion of organic matter into electricity by transferring electrons to a developed circuit . Microorganisms can transfer electrons to the anode electrode in three ways: exogeneous mediators (ones external to the cell) such as potassium ferricyanide, thionine, or nuetral red; using mediators produced by the bacteria; or by direct transfer of electrons from the respiratory enzymes (i.e., cytochromes) to the electrode .

These mediators can divert electrons from the respiratory chain by entering the outer cell membrane, becoming reduced, and then leaving in a reduced state to shuttle the electron to the electrode . Shewanella putrefaciens, Geobacter sulfurreducens, Geobacter metallireducens and Rhodoferax ferrireducens have been shown to generate electricity in a mediatorless MFC .

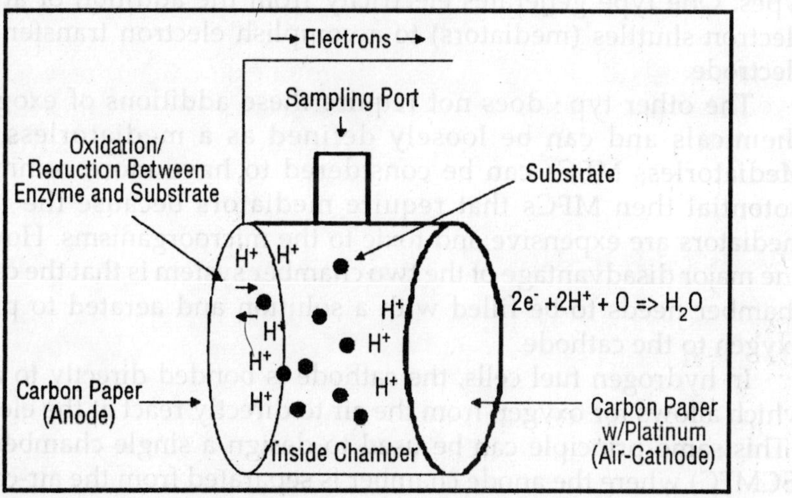

Fig. Representation of Anaerobic (Anode portion) and Aerobic (Cathode portion) Biological Degradation Simultaneous to Electricity Generation in a Single Chamber Microbial Fuel Cell

Bacteria present in mediatorless MFCs have electrochemically active redox enzymes on their outer membranes that transfer the electrons to external materials and therefore, do not require exogeneous chemicals to accomplish electron transfer to the electrode.

When these bacteria oxidize the organic matter present in the wastewater, the electron is shuttled to the electrode and the protons produced diffuse through the water to the counter electrode (cathode) giving this particular electrode a positive characteristic.

Oxygen, the hydrogen protons, and the electron that is connected by a circuit from the anode to the cathode, are then catalytically combined with a platinum catalyst to form water at the cathode . A representation of the biological mechanism previously mentioned is shown within a single chamber MFC.

Note that the mechanism of MFC technology is still in research stages and that many possible reasons for electricity generation that cannot be answered without a better understanding of the characteristics of the electricity generating bacteria in MFCs.

DESIGN STRUCTURES

A typical microbial fuel cell (MFC) consists of two separate chambers which can be inoculated with any type of liquid media. These chambers, an anaerobic anode chamber and an aerobic cathode chamber, are generally separated by a Proton Exchange Membrane (PEM) such as Nafion. A MFC such as this can be classified into two types. One type generates electricity from the addition of artificial electron shuttles (mediators) to accomplish electron transfer to the electrode.

The other type does not require these additions of exogenous chemicals and can be loosely defined as a mediatorless MFC. Mediatorless MFCs can be considered to have more commercial potential then MFCs that require mediators because the typical mediators are expensive and toxic to the microorganisms. However, one major disadvantage of the two chamber system is that the cathode chamber needs to be filled with a solution and aerated to provide oxygen to the cathode.

In hydrogen fuel cells, the cathode is bonded directly to a PEM which allows for oxygen from the air to directly react at the electrode . This same principle can be used to design a single chamber MFC (SCMFC) where the anode chamber is separated from the air-cathode chamber by a gas diffusion layer (GDL) allowing for a passive oxygen transfer to the cathode, eliminating the need for energy intensive air sparging of the liquid. Studies have shown that the MFCs using expensive solid graphite electrodes, but less expensive graphite felt and carbon cloth, can also be used.

PEMs such as Nafion are also expensive and if removed, can substantially reduce the overall cost of the MFC. Platinum is a critical catalyst at the cathode and no alternative metal has been proven to catalyze the combination of oxygen, the hydrogen proton, and the electron in a more efficient manner. These design parameters set a limit on the overall cost reduction of the MFC.

THE MONOD MODEL

The Monod model can describe several important characteristics of microbial growth in a simple periodic culture of microorganisms. This model was first proposed by Nobel Laureate J. Monod more then 50 years ago and is one of the basic models for quantitative microbiology .

This model is typically defined by a differential equation . Many limitations of this model, as well as restrictions of its applications, are well known. With these limitations, many modifications of the model have also been proposed within specific cases. The Monod-type equation used in this analysis is in an algebraic form under the assumption of a steady state condition.

Model Formulation

The Monod model was developed from the Michaelis-Menton equation. The formulation of the Michaelis-Menton equation begins with examining the reaction of an enzyme and a substrate

$$E + S \underset{k_{-1}}{\overset{k_{+1}}{\rightleftharpoons}} E\,S$$

$$E\,S \xrightarrow{k_{+2}} E + P$$

E	=	enzyme
S	=	substrate
$E\,S$	=	enzyme-substrate complex
P	=	product
k_{+1}	=	forward reaction rate constant
k_{-1}	=	reverse reaction rate
k_{+2}	=	reaction rate constant

If the variables e, S, and c are denoted as the concentrations of the total enzyme, substrate, and enzyme-substrate complex, the rate of change of the enzyme-substrate complex, dc/dt , can be expressed as

$$\frac{dc}{dt}=k_{+1}(e-c)S-K_{-1^{c}-K_{+2^{c}}}$$

This equation assumes the value of S is considerably greater then that of e. Assuming a steady state condition, the left hand side of the equation becomes zero and c can be solved for and expressed as

$$c=\frac{eS}{\frac{k_{-1}+k_{+2}}{k_{+1}}+S}$$

From the above enzyme reaction equations, the rate of product formation, v, can be expressed as

$$v=k_{+2^{c}}=\frac{k+2eS}{\frac{k_{-1}+k_{+2}}{k_{+1}}+S}=\frac{VS}{K_s+\frac{K_{+2}}{K_{+1}}+S}=\frac{VS}{K_m+S}$$

$$V \;=\; eK_{+2} \;=\; Maximum\,Rate\,of\,Production$$

$$K_{s} = \frac{k-1}{k+1} \;=\; Equilibrium\,Constant$$

$$K_m \;=\; K_s+\frac{k+2}{k+1}$$

If the value of k+2 is considerably smaller than that of k+1, Km reduces to Ks. When Km = Ks, the Michaelis-Menton equation becomes analogous to the Monod where the Monod-type equation replaces V and v with ¼max (maximum value of specific growth rate) and ¼ (specific growth rate). The Monod-type equation assumes only unicellular growth and can be expressed in the form

$$\mu=\mu\max\left(\frac{S}{K_s+S}\right)$$

The power density produced in a single chamber MFC can be modeled as a function of substrate concentration, S, using an empirical Monod-type equation

$$P=\frac{P_{\max}S}{K_S+S}$$

$$P=Power\,Density$$

$$P_{\max}=Maximum\,Power\,Density$$

$$S=Substrate\,Concentration$$

$$K_s=Half-Saturation\,Constant$$

Parameter Estimation

Linear regression is a parameter estimation optimization problem where Pmax and Ks are the decision variables of the Monod model . The main objective can take one of several forms, all of which attempt to reduce the residuals, thereby improving the model.

A residual is the difference between the observed emperical data to which the model is being fit, and the predicted value output by the model:

$$e_i = y_i - \widehat{yi}$$

$$ei = Residual$$

$$yi = Observed\,Value$$

$$\overline{yi = Predicted\,Value}$$

The most common objectives for this parameter estimation include minimizing the sum of absolute deviations, minimizing the maximum absolute deviation of the residuals, and minimizing the sum of the squared residuals.

WATER QUALITY IMPROVEMENT TECHNOLOGIES

Desalination is a separation process used to reduce the dissolved salt content of saline water to a usable level. All desalination processes involve three liquid streams: the saline feedwater (brackish water or seawater), low-salinity product water, and very saline concentrate (brine or reject water).

The saline feedwater is drawn from oceanic or underground sources. It is separated by the desalination process into the two output streams: the low-salinity product water and very saline concentrate streams. The use of desalination overcomes the paradox faced by many coastal communities, that of having access to a practically inexhaustible supply of saline water but having no way to use it.

Although some substances dissolved in water, such as calcium carbonate, can be removed by chemical treatment, other common constituents, like sodium chloride, require more technically sophisticated methods, collectively known as desalination.

In the past, the difficulty and expense of removing various dissolved salts from water made saline waters an impractical source of potable water. However, starting in the 1950s, desalination began to appear to be economically practical for ordinary use, under certain circumstances.

The product water of the desalination process is generally water with less than 500 mg/1 dissolved solids, which is suitable for most domestic, industrial, and agricultural uses.

A by-product of desalination is brine. Brine is a concentrated salt solution (with more than 35 000 mg/1 dissolved solids) that must be disposed of, generally by discharge into deep saline aquifers or surface waters with a higher salt content. Brine can also be diluted with treated effluent and disposed of by spraying on golf courses and/or other open space areas.

Technical Description

There are two types of membrane process used for desalination: reverse osmosis (RO) and electrodialysis (ED). The latter is not generally used in Latin America and the Caribbean. In the RO process, water from a pressurized saline solution is separated from the dissolved salts by flowing through a water-permeable membrane.

The permeate (the liquid flowing through the membrane) is encouraged to flow through the membrane by the pressure differential created between the pressurized feedwater and the product water, which is at near-atmospheric pressure. The remaining feedwater continues through the pressurized side of the reactor as brine. No heating or phase change takes place.

The major energy requirement is for the initial pressurization of the feedwater. For brackish water desalination the operating pressures range from 250 to 400 psi, and for seawater desalination from 800 to 1 000 psi.

In practice, the feedwater is pumped into a closed container, against the membrane, to pressurize it. As the product water passes through the membrane, the remaining feedwater and brine solution becomes more and more concentrated. To reduce the concentration of dissolved salts remaining, a portion of this concentrated feedwater-brine solution is withdrawn from the container.

Without this discharge, the concentration of dissolved salts in the feedwater would continue to increase, requiring ever-increasing energy inputs to overcome the naturally increased osmotic pressure.

A reverse osmosis system consists of four major components/ processes:

- Pretreatment
- Pressurization
- Membrane separation and
- Post-treatment stabilization. Illustrates the basic components of a reverse osmosis system.

Pretreatment: The incoming feedwater is pretreated to be compatible with the membranes by removing suspended solids, adjusting the pH, and adding a threshold inhibitor to control scaling caused by constituents such as calcium sulphate.

Pressurization: The pump raises the pressure of the pretreated feedwater to an operating pressure appropriate for the membrane and the salinity of the feedwater.

Separation: The permeable membranes inhibit the passage of dissolved salts while permitting the desalinated product water to pass through. Applying feedwater to the membrane assembly results in a freshwater product stream and a concentrated brine reject stream. Because no membrane is perfect in its rejection of dissolved salts, a small percentage of salt passes through the membrane and remains in the product water.

Reverse osmosis membranes come in a variety of configurations. Two of the most popular are spiral wound and hollow fine fibre membranes They are generally made of cellulose acetate, aromatic polyamides, or, nowadays, thin film polymer composites. Both types are used for brackish water and seawater desalination, although the specific membrane and the construction of the pressure vessel vary according to the different operating pressures used for the two types of feedwater.

Stabilization: The product water from the membrane assembly usually requires pH adjustment and degasification before being transferred to the distribution system for use as drinking water. The product passes through an aeration column in which the pH is elevated from a value of approximately 5 to a value close to 7. In many cases, this water is discharged to a storage cistern for later use.

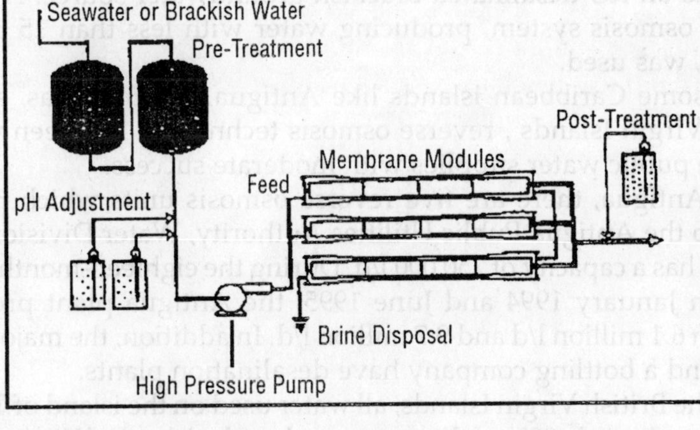

Fig. Elements of the Reverse Osmosis Desalination Process.

Extent of Use

The capacity of reverse osmosis desalination plants sold or

installed during the 20-year period between 1960 and 1980 was 1 050 600 m^3/day. During the last 15 years, this capacity has continued to increase as a result of cost reductions and technological advances. RO-desalinated water has been used as potable water and for industrial and agricultural purposes.

Potable Water Use: RO technology is currently being used in Argentina and the northeast region of Brazil to desalinate groundwater. New membranes are being designed to operate at higher pressures (7 to 8.5 atm) and with greater efficiencies (removing 60% to 75% of the salt plus nearly all organics, viruses, bacteria, and other chemical pollutants).

Industrial Use: Industrial applications that require pure water, such as the manufacture of electronic parts, speciality foods, and pharmaceuticals, use reverse osmosis as an element of the production process, where the concentration and/or fractionating of a wet process stream is needed.

Agricultural Use: Greenhouse and hydroponic farmers are beginning to use reverse osmosis to desalinate and purify irrigation water for greenhouse use (the RO product water tends to be lower in bacteria and nematodes, which also helps to control plant diseases).

Reverse osmosis technology has been used for this type of application by a farmer in the State of Florida, U.S.A., whose production of European cucumbers in a 22 ac. greenhouse increased from about 4 000 dozen cucumbers/day to 7 000 dozen when the farmer changed the irrigation water supply from a contaminated surface water canal source to an RO-desalinated brackish groundwater source. A 300 l/d reverse osmosis system, producing water with less than 15 mg/1 of sodium, was used.

In some Caribbean islands like Antigua, the Bahamas, and the British Virgin Islands , reverse osmosis technology has been used to provide public water supplies with moderate success.

In Antigua, there are five reverse osmosis units which provide water to the Antigua Public Utilities Authority, Water Division. Each RO unit has a capacity of 750 000 l/d. During the eighteen-month period between January 1994 and June 1995, the Antigua plant produced between 6.1 million l/d and 9.7 million l/d. In addition, the major resort hotels and a bottling company have desalination plants.

In the British Virgin Islands, all water used on the island of Tortola, and approximately 90% of the water used on the island of Virgin Gorda, is supplied by desalination. On Tortola, there are about 4 000 water connections serving a population of 13 500 year-round residents and approximately 256 000 visitors annually. In 1994, the government water utility bought 950 million liters of desalinated water for distribution on Tortola.

On Virgin Gorda, there are two seawater desalination plants. Both have open seawater intakes extending about 450 m offshore. These plants serve a population of 2 500 year-round residents and a visitor population of 49 000, annually. There are 675 connections to the public water system on Virgin Gorda. In 1994, the government water utility purchased 80 million liters of water for distribution on Virgin Gorda.

In South America, particularly in the rural areas of Argentina, Brazil, and northern Chile, reverse osmosis desalination has been used on a smaller scale.

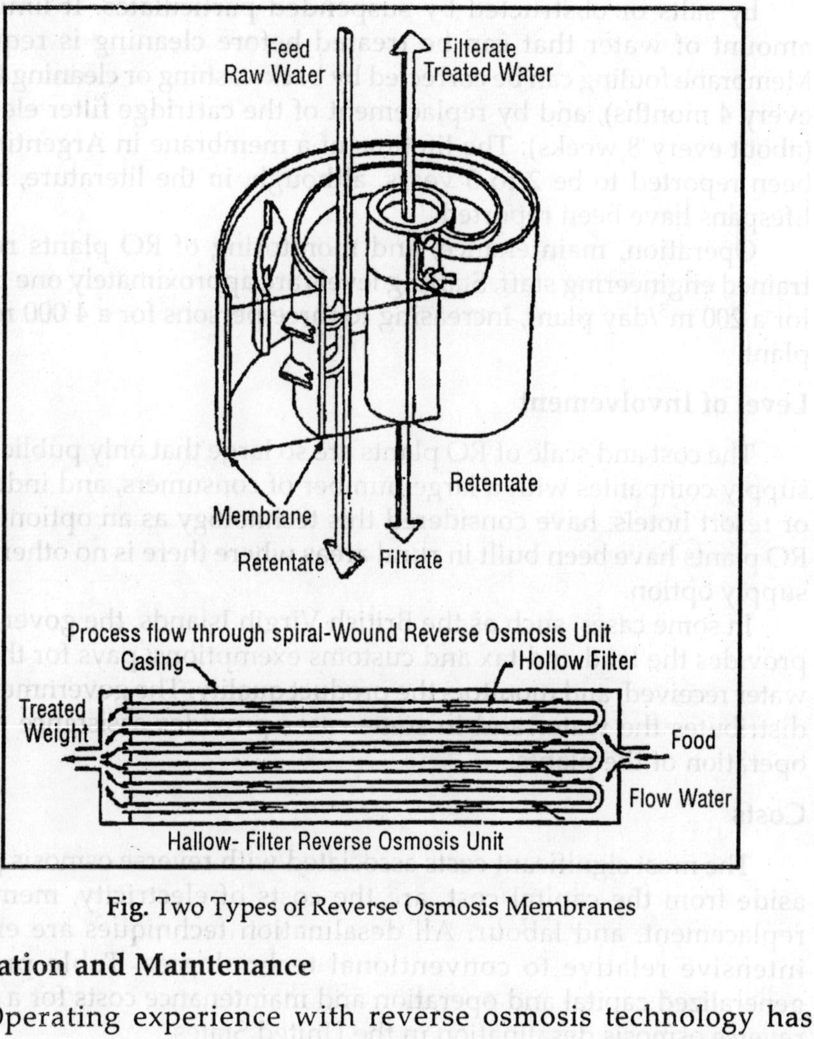

Fig. Two Types of Reverse Osmosis Membranes

Operation and Maintenance

Operating experience with reverse osmosis technology has

improved over the past 15 years. Fewer plants have had long-term operational problems. Assuming that a properly designed and constructed unit is installed, the major operational elements associated with the use of RO technology will be the day-to-day monitoring of the system and a systematic program of preventive maintenance.

Preventive maintenance includes instrument calibration, pump adjustment, chemical feed inspection and adjustment, leak detection and repair, and structural repair of the system on a planned schedule.

The main operational concern related to the use of reverse osmosis units is fouling. Fouling is caused when membrane pores are clogged by salts or obstructed by suspended particulates. It limits the amount of water that can be treated before cleaning is required. Membrane fouling can be corrected by backwashing or cleaning (about every 4 months), and by replacement of the cartridge filter elements (about every 8 weeks). The lifetime of a membrane in Argentina has been reported to be 2 to 3 years, although, in the literature, higher lifespans have been reported.

Operation, maintenance, and monitoring of RO plants require trained engineering staff. Staffing levels are approximately one person for a 200 m^3/day plant, increasing to three persons for a 4 000 m^3/day plant.

Level of Involvement

The cost and scale of RO plants are so large that only public water supply companies with a large number of consumers, and industries or resort hotels, have considered this technology as an option. Small RO plants have been built in rural areas where there is no other water supply option.

In some cases, such as the British Virgin Islands, the government provides the land and tax and customs exemptions, pays for the bulk water received, and monitors the product quality. The government also distributes the water and in some cases provides assistance for the operation of the plants.

Costs

The most significant costs associated with reverse osmosis plants, aside from the capital cost, are the costs of electricity, membrane replacement, and labour. All desalination techniques are energy-intensive relative to conventional technologies. Table presents generalized capital and operation and maintenance costs for a 5 mgd reverse osmosis desalination in the United States.

Reported cost estimates for RO installations in Latin American and the Caribbean. The variation in these costs reflects site-specific factors such as plant capacity and the salt content of the feedwater.

The International Desalination Association (IDA) has designed a Seawater Desalting Costs Software Program to provide the mathematical tools necessary to estimate comparative capital and total costs for each of the seawater desalination processes.

Table. U.S. Army Corps of Engineers Cost Estimates for RO Desalination Plants in Florida

Feedwater Type	Capital Cost per Unit of Daily Capacity ($/m³/day)	Operation & Maintenance per Unit of Production ($/m³)
Brackish water	380 - 562	0.28 - 0.41
Seawater	1341 - 2379	1.02 - 1.54

Table. Comparative Costs of RO Desalination for Several Latin American and Caribbean Developing Countries

Country	Capital Cost ($/m³/day)	Operation and Maintenance ($/m³)	Production Cost* ($/m³)[a]
Antigua	264 - 528	0.79 - 1.59	
Argentina		3.25	
Bahamas			4.60 - 5.10
Brazil	1454 - 4483		0.12 - 0.37
British Virgin Islands	1190 - 2642		[b]3.40 - 4.30
Chile	1300		1.00

Effectiveness of the Technology

Twenty-five years ago, researchers were struggling to separate product waters from 90% of the salt in feedwater at total dissolved solids (TDS) levels of 1 500 mg/1, using pressures of 600 psi and a flux through the membrane of 18 l/m²/day. Today, typical brackish installations can separate 98% of the salt from feedwater at TDS levels of 2 500 to 3 000 mg/1, using pressures of 13.6 to 17 atm and a flux of 24 l/m^2/day - and guaranteeing to do it for 5 years without having to replace the membrane.

Today's state-of-the-art technology uses thin film composite membranes in place of the older cellulose acetate and polyamide membranes.

The composite membranes work over a wider range of pH, at higher temperatures, and within broader chemical limits, enabling them to withstand more operational abuse and conditions more commonly found in most industrial applications.

In general, the recovery efficiency of RO desalination plants increases with time as long as there is no fouling of the membrane.

Suitability

This technology is suitable for use in regions where seawater or brackish groundwater is readily available.

Advantages

- The processing system is simple; the only complicating factor is finding or producing a clean supply of feedwater to minimize the need for frequent cleaning of the membrane.
- Systems may be assembled from prepackaged modules to produce a supply of product water ranging from a few liters per day to 750 000 l/day for brackish water, and to 400 000 l/day for seawater; the modular system allows for high mobility, making RO plants ideal for emergency water supply use.
- Installation costs are low.
- RO plants have a very high space/production capacity ratio, ranging from 25 000 to 60 000 l/day/m^2.
- Low maintenance, nonmetallic materials are used in construction.
- Energy use to process brackish water ranges from 1 to 3 kWh per 1 0001 of product water.
- RO technologies can make use of use an almost unlimited and reliable water source, the sea.
- RO technologies can be used to remove organic and inorganic contaminants.
- Aside from the need to dispose of the brine, RO has a negligible environmental impact.
- The technology makes minimal use of chemicals.

Disadvantages

- The membranes are sensitive to abuse.
- The feedwater usually needs to be pretreated to remove particulates (in order to prolong membrane life).
- There may be interruptions of service during stormy weather (which may increase particulate resuspension and the amount of suspended solids in the feedwater) for plants that use seawater.
- Operation of a RO plant requires a high quality standard for materials and equipment.
- There is often a need for foreign assistance to design, construct, and operate plants.
- An extensive spare parts inventory must be maintained, especially if the plants are of foreign manufacture.
- Brine must be carefully disposed of to avoid deleterious environmental impacts.
- There is a risk of bacterial contamination of the membranes; while bacteria are retained in the brine stream, bacterial growth on the membrane itself can introduce tastes and odors into the product water.
- RO technologies require a reliable energy source.
- Desalination technologies have a high cost when compared to other methods, such as groundwater extraction or rainwater harvesting.

Cultural Acceptability

RO technologies are perceived to be expensive and complex, a perception that restricts them to high-value coastal areas and limited use in areas with saline groundwater that lack access to more conventional technologies. At this time, use of RO technologies is not widespread.

Further Development of the Technology

The seawater and brackish water reverse osmosis process would be further improved with the following advances:

- Development of membranes that are less prone to fouling, operate at lower pressures, and require less pretreatment of the feedwater.
- Development of more energy-efficient technologies that are simpler to operate than the existing technology; alternatively, development of energy recovery methodologies that will make better use of the energy inputs to the systems.

- Commercialization of the prototype centrifugal reverse osmosis desalination plant developed by the Canadian Department of National Defence; this process appears to be more reliable and efficient than existing technologies and to be economically attractive.

DESALINATION BY DISTILLATION

Distillation is the oldest and most commonly used method of desalination. The world's first land-based desalination plant, a multiple-effect distillation (MED) process plant that had a capacity of 60 m^3/day, was installed on Curaçao, Netherlands Antilles, in 1928. Further commercial development of land-based seawater distillation units took place in the late 1950s, and initially relied on the technology developed for industrial evaporators (such as sugar concentrators) and for the shipboard distillation plants which were built during World War n.

The multistage-flash (MSF), MED, and vapour-compression (VC) processes have led to the widespread use of distillation to desalinate seawater.

Technical Description

Distillation is a phase separation method whereby saline water is heated to produce water vapour, which is then condensed to produce freshwater. The various distillation processes used to produce potable water, including MSF, MED, VC, and waste-heat evaporators, all generally operate on the principle of reducing the vapour pressure of water within the unit to permit boiling to occur at lower temperatures, without the use of additional heat.

Distillation units routinely use designs that conserve as much thermal energy as possible by interchanging the heat of condensation and heat of vaporization within the units. The major energy requirement in the distillation process thus becomes providing the heat for vaporization to the feedwater.

Multistage Flash (MSF)

Shows a simplified schematic of a multistage-flash unit. The incoming seawater passes through the heating stage(s) and is heated further in the heat recovery sections of each subsequent stage. After passing through the last heat recovery section, and before entering the first stage where flash-boiling (or flashing) occurs, the feedwater is further heated in the brine heater using externally supplied steam.

This raises the feedwater to its highest temperature, after which it is passed through the various stages where flashing takes place. The vapour pressure in each of these stages is controlled so that the heated brine enters each chamber at the proper temperature and pressure (each lower than the preceding stage) to cause instantaneous and violent boiling/evaporation.

The freshwater is formed by condensation of the water vapour, which is collected at each stage and passed on from stage to stage in parallel with the brine. At each stage, the product water is also flash-boiled so that it can be cooled and the surplus heat recovered for preheating the feedwater.

Because of the large amount of flashing brine required in an MSF plant, a portion (50% to 75%) of the brine from the last stage is often mixed with the incoming feedwater, recirculated through the heat recovery sections of the brine heater, and flashed again through all of the subsequent stages. A facility of this type is often referred to as a "brine recycle" plant.

This mode of operation reduces the amount of water-conditioning chemicals that must be added, and can significantly affect operating costs.

On the other hand, it increases the salinity of the brine at the product end of the plant, raises the boiling point, and increases the danger of corrosion and scaling in the plant.

In order to maintain a proper brine density in the system, a portion of the concentrated brine from the last stage is discharged to the ocean. The discharge flow rate is controlled by the brine concentration at the last stage.

Multiple Effect (MED)

In multiple-effect units steam is condensed on one side of a tube wall while saline water is evaporated on the other side (in a manner similar to the VC process shown.The energy used for evaporation is the heat of condensation of the steam. Usually there is a series of condensation-evaporation processes taking place (each being an "effect").

The saline water is usually applied to the tubes in the form of a thin film so that it will evaporate easily. Although this is an older technology than the MSF , it has not been extensively utilized for water production. However, a new type of low-temperature, horizontal-tube MED process has been successfully developed and used in the Caribbean. These plants appear to be very rugged, easy to operate, and economical, since they can be made of aluminum or other low-cost materials.

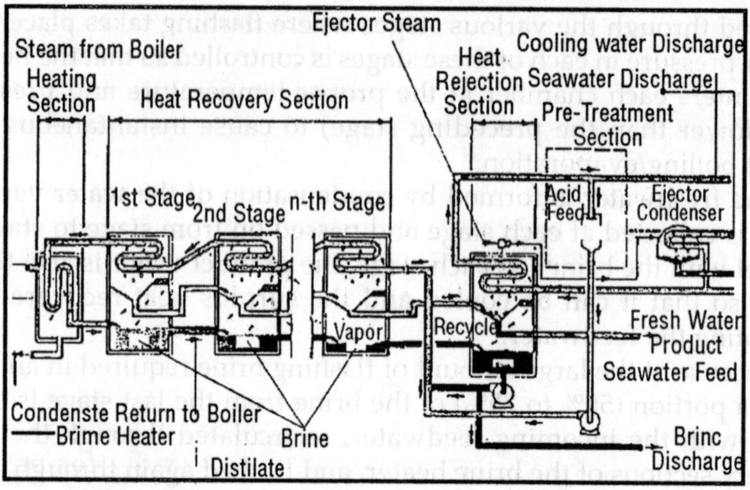

Fig. Simplified Schematic of a Multistage Flash (MSF) Distillation Plant

Vapour Compression (VC)

The vapour-compression process uses mechanical energy rather than direct heat as a source of thermal energy. Water vapour is drawn from the evaporation chamber by a compressor and except in the first stage is condensed on the outsides of tubes in the same chambers, as is shown in.

The heat of condensation is used to evaporate a film of saline water applied to the insides of the tubes within the evaporation chambers. These units are usually built with capacities of less than 100 m'/day and are often used at resorts and industrial sites.

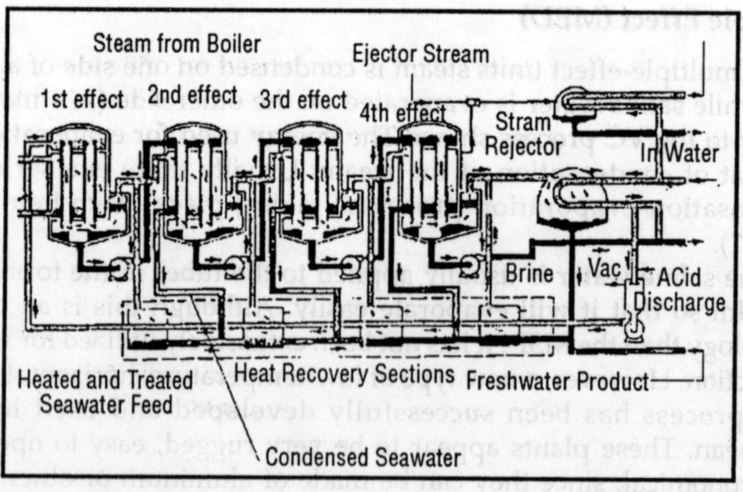

Fig. Simplified Schematic of a Vapour Compression Distillation Plant.

Membrane Distillation

Membrane distillation is a relatively new process, having been introduced commercially only in the last few years. The process works by using a specialized membrane which will pass water vapour but not liquid water. This membrane is placed over a moving stream of warm water, and as the water vapour passes through the membrane it is condensed on a second surface which is at a lower temperature than that of the feedwater.

Dual Purpose

Most of the large distillation units in the world are dual-purpose facilities. Specifically, they derive their source of thermal energy from steam that has been used for other purposes, usually for power generation. Thus, the feedwater is heated in a boiler to a high energy level and passed through a steam turbine before the steam is extracted for use at a lower temperature to provide the heat required in the distillation plants.

Extent of Use

Since 1971, about 65 single-purpose service or experimental plants have been installed in Latin America and the Caribbean, with capacities ranging from 15 to 1 000 m^3/day. In Mexico they supply freshwater to fishing villages and/or tourist resorts in Baja California and in the north-central and southeastern parts of the country. They also provide freshwater to agricultural communities.

Desalination for municipal freshwater supply purposes started in Mexico in the late 1960s, when the Federal Electricity Commission installed two 14 000 m^3/day MSF distillation units in its Rosarito Power Plant in the city of Tijuana in northwest Mexico. At that time, those units were among the largest in the world.

The Federal Electricity Commission currently operates about 31 desalination plants to produce high-quality boiler make-up water, and maintains the two dual-purpose units in Tijuana. The Mexican Navy also installed some smaller solar distillation plants to provide a supply of freshwater to some islands in the Pacific Ocean.

PEMEX, the national oil company of Mexico, operates about 62 small seawater desalination plants for human freshwater consumption on off-shore oil platforms or ships. These distillation units are mainly VC, waste heat, submerged-tube evaporators, and RO plants.

The island of Curaçao, in the Netherlands Antilles, currently has two distillation plants. One is for public water supply and the other is used by the oil refinery PEDEVESA. Both use the MSF process. The public supply plant has a maximum design capacity of 47 000 m^3/day (although the average daily production is currently 41 000 m^3/day), which is higher than the estimated domestic water consumption of 35 000 m^3/day.

Operation and Maintenance

Most plants are installed in isolated locations where construction is troublesome and where the availability of fuel, chemicals, and spare parts is limited. In these places, there is usually also a scarcity of qualified personnel; therefore, people are often selected from the local communities and trained to operate the plants.

The operation of distillation plants requires careful planning, well-trained operators, and adequate operation and maintenance budgets to guarantee the supply of good quality water. Except for an annual shutdown of 6 to 8 weeks for general inspection and maintenance, the operation of desalination plants is usually continuous. Maintenance and preventive maintenance work, for a MSF plant, consists of:

- Repairing damage (cracks) to the stainless steel liners in the stages.
- Removing scale and marine growths in the tubes in all stages using high pressure "hydrolaser" sprayers.
- Removing the vacuum system ejectors for cleaning, inspection, and replacement as necessary; most parts have a lifetime of 3 to 4 years.
- Inspecting all pumps and motors, replacing bearings and bushings, and renewing protective coatings on exposed parts (e.g., pumps must be primed and painted before being installed).

Level of Involvement

The manufacturing capacity to produce MSF evaporators is available in those places where power plant equipment is fabricated. Thus, many countries in Latin America have the potential to manufacture locally the equipment needed to develop desalination plants.

Further, some local manufacturers have signed licensing agreements with major foreign desalination manufacturing firms as a result of governmental policies of import substitution, in order to offer desalination equipment, particularly MSF plants, to the electric-generating industry in the region.

In die Caribbean, desalination by distillation is being used primarily in the private sector, especially in the tourist industry. Some industrial concerns and power companies have incorporated distillation into their operations as part of a dual process approach. Government participation has been very limited.

Future developments of this technology, which are expected to reduce the cost of desalination plants, will be likely to encourage greater government participation in the use of distillation in the development of public water supply systems.

Costs

The production cost of water is a function of the type of distillation process used, the plant capacity, the salinity in the feedwater (seawater or brackish water), and the level of familiarity with the distillation process that exists in the region. Table shows a range of costs that have been reported by different countries using this technology.

Production costs appear to increase in proportion to the capacity of the plant. In many applications, distillation provides the best means of achieving waters of high purity for industrial use: for volumes of less than 4 000 m'/day, the VC process is likely to be most effective; above that range, the MSF process will probably be preferable.

Table. Estimated Cost of Distillation Processes in Latin American Countries

Country	Distillation Process	Capital Cost ($)	Operation and Maintenance Cost ($/year)	Energy Cost ($/year)	Production Cost ($/m³/year)
Aruba	MSF	10000	1612	2860	
Chile	VC				1.47
U.S. Virgin Is.	MED and VC				4.62
Curaçao	MSF				4.31

Effectiveness of the Technology

Desalination of seawater is a relatively expensive method of obtaining freshwater. The MSF system has proved to be a very efficient system, when properly maintained. It produces high quality product water (between 2 and 150 mg/1 of total dissolved solids at the plant in Curaçao); TDS contents of less than 10 mg/1 have been reported from the VC plant in Chile. Because the water is boiled, the risk of bacterial or pathogenic virus contamination of the product water is minimal.

Suitability

MSF plants have been extensively used in the Middle East, North Africa, and the Caribbean. Although MED is an older technology than the MSF process, having been used in sugar refineries, it has not been extensively utilized for water production.

However, the new low-temperature horizontal-tube MED process has been successfully used in the Caribbean, usually in units with capacities of less than 100 m'/d (25,000 gpd) installed at resorts and industrial sites.

Advantages

- Distillation offers significant savings in operational and maintenance costs compared with other desalination technologies.
- In most cases, distillation does not require the addition of chemicals or water softening agents to pretreat feedwater.
- Low temperature distillation plants are energy-efficient and cost-effective to operate.
- Many plants are fully automated and require a limited number of personnel to operate.
- Distillation has minimal environmental impacts, although brine disposal must be considered in the plant design.
- The technology produces high-quality water, in some cases having less than 10 mg/1 of total dissolved solids.
- Distillation can be combined with other processes, such as using heat energy from an electric-power generation plant.

Disadvantages

- Some distillation processes are energy-intensive, particularly the large-capacity plants. «Disposal of the brine is a problem in many regions.
- The distillation process, particularly MSF distillation, is very costly.
- Distillation requires a high level of technical knowledge to design and operate.
- The technology requires the use of chemical products, such as acids, that need special handling.

CULTURAL ACCEPTABILITY

Despite significant progress toward becoming more energy-efficient and cost-effective, the level of community acceptance of distillation technologies is still limited. Their use is mainly restricted to resort hotels and high-value-added industries, and to the Caribbean islands.

Further Development of the Technology

Research into the falling (or spray) film MED thermal desalination process suggests that further development of distillation technologies can produce product waters that are comparable in quality to those produced with current MSF technologies and also offer additional advantages, including lower pumping requirements, higher heat transfer rates, and greatly reduced pressure differentials across the heat transfer surfaces.

These favorable comparisons also apply to a falling (or spray) film VC design. Some additional considerations include:

- Lower operating temperatures (150 to 180° F)(66 to 82° C) and vapour velocities, reducing system losses.
- Higher thermal efficiencies to reduce fuel and energy costs.
- Improved materials for evaporator heat transfer surfaces (aluminum has two major benefits over other materials: a lower cost than copper-nickel, with nearly triple the thermal conductivity and higher operating temperatures, with an upper limit of 150° F (63" C) for aluminum alloys containing approximately 2% magnesium).
- Improved coatings for use in shell construction (with aluminum evaporator heat transfer surfaces, it is essential to prevent corrosion caused by the proximity of other metal ions; the carbon steel shell must be appropriately coated, and provision made for all supporting structures to be protected).
- Improved piping material for use with low temperature distillation techniques; piping should be of PVC, fiberglass, or other suitable non-metallic material.

A further alternative and promising new concept for a dual purpose plant has been the development of an evaporative condenser which is equipped with dimpled flat plate elements that could greatly increase the efficiency of this type of plant.

DISINFECTION BY BOILING AND CHLORINATION

Boiling and chlorination are the most common water and wastewater disinfection processes in use throughout the world. Boiling is primarily used in rural areas in developing countries to eliminate living organisms, especially bacteria, present in the water.

It is also used in emergencies when other, more sophisticated methods of disinfection are not available. Prior to the development of chlorination, boiling was the principal method used to kill pathogenic organisms.

TECHNICAL DESCRIPTION

Boiling

Boiling is a very simple method of water disinfection. Heating water to a high temperature, 100°C, kills most of the pathogenic organisms, particularly viruses and bacteria causing waterborne diseases. In order for boiling to be most effective, the water must boil for at least 20 minutes.

Since boiling requires a source of heat, rudimentary or non-conventional methods of heat generation may be needed in areas where electricity or fossil fuels are not available.

Chlorination

Chlorination has become the most common type of wastewater and water disinfection. It should be noted that it is designed to kill harmful organisms, and generally does not result in sterile water (free of all microorganisms). Two types of processes are generally used: hypochlorination, employing a chemical feed pump to inject a calcium or sodium hypochlorite solution, and gas chlorination, using compressed chlorine gas.

Hypochlorination

Calcium hypochlorite is available commercially in either a dry or wet form. High-test calcium hypochlorite (HTH), the form most frequently used, contains about 60% available chlorine. Because calcium hypochlorite granules or pellets are readily soluble in water and are relatively stable under proper storage conditions, they are often favored over other forms. Shows a typical hypochlorite installation.

Sodium hypochlorite is available in strengths from 1.5% to 15%, with 3% available chlorine as the typical strength used in water treatment applications. The higher the strength of the chlorine solution, the more rapidly it decomposes and the more readily it is degraded by exposure to light and heat.

It must therefore be stored in a cool location and in a corrosion-resistant tank. Typically, 30 minutes of chlorine contact time is required for optimal disinfection with good mixing. Water supply treatment dosages are established on the basis of maintaining a residual concentration of chlorine in the treated water.

Water-based solutions of either the liquid or the dry form of hypochlorite are prepared in predetermined stock solution strengths. Solutions are injected into the water supply using special chemical metering pumps called hypochlorinators. Positive displacement types are the most accurate and reliable and are commonly preferred to hypochlorinators employing other feed principles (usually based on suction). Positive-displacement-type hypochlorinators are readily available at relatively modest costs.

These small chemical-feed pumps are designed to pump (inject under pressure) an aqueous solution of chlorine into the water system. They are designed to operate against pressures as high as 100 psi, but may also be used to inject chlorine solutions under ambient (atmospheric) or negative head conditions.

Hypochlorinators come in various capacities ranging from 3.8 to 227 l/day. Usually, the pumping rate is manually adjusted by varying the stroke of the pump's piston or diaphragm.

Once the stroke is set, the hypochlorinator accurately feeds chlorine into the system at that rate, maintaining a constant dose. This works well if the water supply rate and the output of the pump are fairly constant.

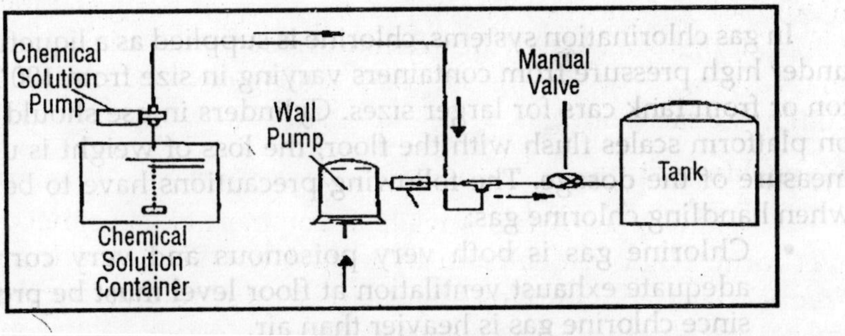

Fig. A Typical Hypochlorite Installation

Montserrat has been using floating chlorinators, but in response to concern expressed by the Director of Health Services that they leave chlorine residues in the water supply and that "the chlorine values are generally too low to guarantee safety," the Montserrat Water Authority looked in to various other methods and decided on gaseous chlorine. It is now proceeding cautiously to replace floating chlorinators with gas chlorination as treatment plant operators are trained in the new system.

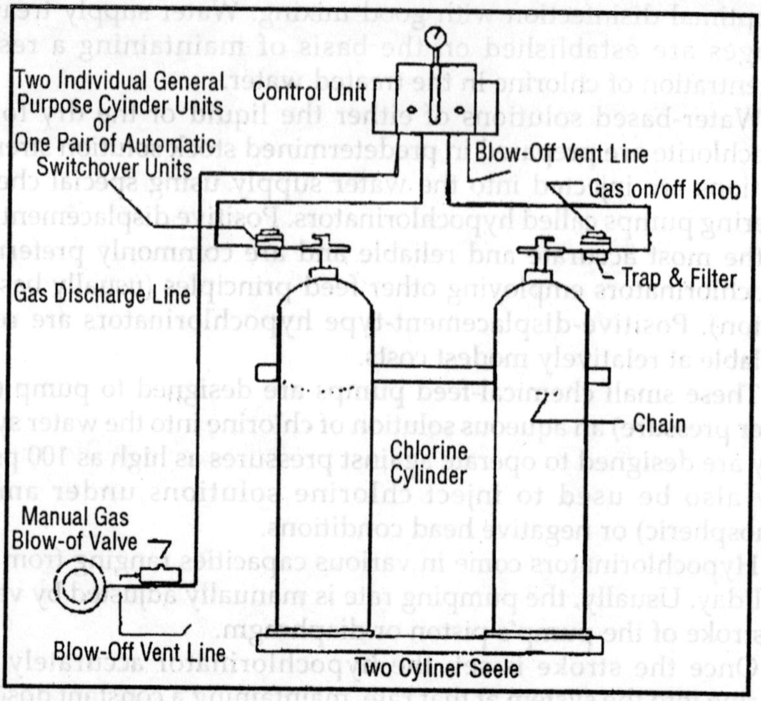

Fig. A Typical Chlorine Cylinder Setup for Gas Chlorination Treatment

Gas Chlorination

In gas chlorination systems, chlorine is supplied as a liquefied gas under high pressure from containers varying in size from 100 Ib to 1 ton or from tank cars for larger sizes. Cylinders in use should be set on platform scales flush with the floor; the loss of weight is used as measure of the dosage. The following precautions have to be taken when handling chlorine gas:

- Chlorine gas is both very poisonous and very corrosive; adequate exhaust ventilation at floor level must be provided since chlorine gas is heavier than air.
- Chlorine-containing liquids and gases can be handled in wrought-iron piping; however, chlorine solutions are highly corrosive and should be handled in rubber-lined or corrosion-resistant plastic piping with hard rubber fittings where necessary.
- Pressurized chlorine gas should never be piped in silver, glass, Teflon, or other piping material that cannot handle the pressure; exposure to concentrated chlorine gas can be fatal.

A gas chlorinator meters the gas flow and mixes the gas with water.

The resulting chlorine solution is then injected into the product water. Small water supplies can be effectively served by a 100 or 150 lb container; larger containers are not recommended for small systems, as they require special hoists and cradles.

(Chlorine gas is a highly toxic lung irritant compound and special facilities are required for storing and housing gas chlorinators.) The advantage of this method, however, is the convenience afforded by the relatively large quantity of chlorine gas available for continuous periods of operation lasting several days or weeks, without the need to mix chemicals. Shows a typical chlorine gas cylinder system for gas chlorination treatment.

Extent of Use

Boiling is a primary technology used to control the spread of waterborne diseases. It is a traditional technology that was used prior to the advent of existing technologies. It is still used in areas where the energy supplies and modem facilities needed for other technologies are lacking, and in areas where the quality of the water supply is questionable.

The most common system of disinfection in Latin America and the Caribbean is chlorination. Chlorine tablets, liquid, powder, and gas are widely used. Chlorination of water supplies on an emergency basis was practiced in the region as early as about 1850. At present, chlorination of both water supplies and wastewater is widespread. Chlorination for disinfection is used to prevent the spread of waterborne diseases and to control algal growth and odors. Economics, ease of operation, and convenience are the main factors used to evaluate disinfection processes.

For safety, and to ensure a constant supply of chlorine, on-site generation is recommended. Most commercially available chlorine generation equipment will operate on waters ranging in salinity from freshwater to seawater, and also on brine solutions prepared for the purpose.

Hypochlorite solutions prepared from seawater are usually limited to about 1 800 mg/l of available chlorine, and those produced from brine to about 8 000 mg/l. Heavy metal ions present in seawater interfere with the stability of hypochlorite solutions prepared using water from this source.

Operation and Maintenance

Gas Chlorinators. Gas chlorinators have an advantage in situations where water flow rates are variable, because the chlorine feed rates may be synchronized to inject variable quantities of chlorine into the product water.

Capital costs of gas chlorination, however, are somewhat greater, but chemical costs may be less. Normal operation of a gas chlorinator requires routine observation and preventive maintenance. Daily duties of an operator should include the following tasks:

- Reading the chlorinator rotameter daily and recording the information.
- Reading the product water flow meters and recording the amount of water pumped.
- Checking the chlorine residual levels in the distribution system and, as necessary, adjusting the rotameter to increase the feed rate if they are too low and decrease it if they are too high.
- Calculating the chlorine usage, and ordering further chlorine stocks if necessary.
- Cleaning the equipment and the building weekly, cleaning the "Y" strainer three times a week, and replacing the gaskets periodically.
- Performing preventive maintenance on the equipment.

Hypochlorinators

Because of its oxidizing potential, calcium hypochlorite should be stored in a cool, dry location, away from other chemicals, in corrosion-resistant containers. Operators should perform the following maintenance tasks:

- Reading and recording the level of the solution tank at the same time every day.
- Reading the product water flow meters and recording the amount of water pumped.
- Checking the chlorine residual levels in the system and adjusting the chlorine feed rate as necessary, in order to maintain a chlorine residual level of 0.2 mg/l at the most remote point in the distribution system (the suggested free chlorine residual for treated water or well water is 0.5 mg/l at the point of chlorine application, provided that the 0.2 mg/l concentration is maintained throughout the distribution system). The chlorine feed rate of a floating chlorinator must be adjusted daily to increase or decrease the dosage in conformity with the water output of the treatment plant.
- Checking and adjusting the chemical feed pump operation; most hypochlorinators have a dial indicating the chlorine feed

rate, with a range from 0 to 10, the pointer of which should initially be set to approximately 6 or 7, when using a 2 % hypochlorite solution. The pump should be operated in the upper ranges of the dial to ensure that the strokes or pulses from the pump are frequent enough so that the chlorine will be fed continuously into the water being treated.

- Replacing the chemicals and washing the chemical storage tank as necessary so that a 15-to 30-day supply of chlorine is on hand to meet future needs; hypochlorite solutions, however, should be prepared only in quantities needed for two to three days of operation, in order to preserve their potency.
- Checking the operation of the check valve.
- Inspecting and cleaning the feeder valves. Commercial sodium hypochlorite solutions (such as Clorox) contain an excess of caustic soda (sodium hydroxide, NaOH); when diluted with highly alkaline water, they produce a solution that is supersaturated with calcium carbonate, which tends to form a coating on the valves in the solution feeder. Similarly, in systems using calcium hypochlorite (HTH), when sodium fluoride is injected at the same point as the hypochlorite solution the calcium and fluoride ions combine and form a coating. The coated valves will not seat properly and the feeder will fail to chlorinate the product water properly. (Small hypochlorinators are sealed so that they cannot be repaired without replacing the entire unit. Otherwise, they require very little maintenance, mostly consisting of a periodic oil change and lubrication.)

Frequent visits are required to the chlorination points in the distribution system to make adjustments, to clear PVC tubing of sludge formation that stops tablets from dissolving, and to recharge tablets.

Level of Involvement

Boiling is exclusively the responsibility of individual users. Chlorination is normally conducted by the private sector in small-scale hypochlorite treatment systems. Regional or large-scale systems require the involvement of a public utility or regional water supply authority, particularly if gas chlorination is used. For large systems, government involvement and financing are required.

Costs

The cost of boiling is related to the cost of the energy used in the process. The cost of chlorination systems varies considerably depending on the geographic location and the type of chlorination system used. Table shows a comparison of capital costs of two different chlorination systems.

Table. Comparison of Capital Costs of Chlorination Systems ($)

Item	*Gas Chlorination*	*Hypochlorite Tablets*
Equipment	10482	875
Installation	1516	150
Building	10000	-
Total	21999	1020

Effectiveness of the Technology

Boiling is a very effective disinfection technology, but it is recommended only as a backup to other technologies because of its volume limitations and energy requirements.

Chlorination is a very effective and well-known technology. Its effectiveness is a function of the quality of the water that is being chlorinated and the method of chlorination used. Normally gas chlorination is a more efficient method of disinfection, although a system based on the use of hypochlorite tablets is easier to operate and maintain and is preferred by individual users. Table shows a comparison of the two methods as used on the Caribbean island of Montserrat.

Table. Technological Efficiency of Chlorination Methods

Chlorination Method	Tablets/Granules	Chlorine Gas
Chlorine usage	201 lb Cl^2	102 lb gas
Total Cl residue	27.1 mg/l	40.5 mg/l
Residue/Cl_2 ratio	0.13mg/lb Cl^2	0.46 mg/lb of Cl^2
% of available chlorine	65%	100%
Treatment cost	$1532	$172

Suitability

Boiling is applicable everywhere, although it is now most often used in emergencies or in rural areas where chlorinated public water supplies are not available. Chlorination can be used in most areas depending on the availability of chemicals. Gas Chlorination, however, is best used in controlled situations such as provided by a public water utility.

Advantages

As was noted above, boiling, while an effective technology, is generally considered to be a secondary or emergency means of disinfecting water supplies. For this reason, the following advantages refer to Chlorination systems:

- The systems are extremely reliable; the hypochlorite system is somewhat easier to operate than the gas system because the operators need not be as skilled or as cautious.
- Chlorination is less costly than other disinfection systems and is generally easier to implement; chlorine (Cl_2) can be made in the region and safety considerations for its production, transportation, and use are well known.

Hypochlorinator System

- Hypochlorite compounds are non-flammable.
- Hypochlorite does not present the same hazards as gaseous chlorine and therefore is safer to handle; spills may be cleaned up with large volumes of water.
- Floating chlorinators can be adapted to small community systems or individual rainwater collector systems. They easy to construct and to transport. However, they cannot easily guarantee uniform residual chlorine concentrations.

Gas Feeder System

- Gas feeder systems are fitted with valves to automatically close the vacuum regulator in case of leaks or accidental breaks in the vacuum line, stopping gas flow at source.
- The systems have an automatic shut-off in case of interruption of feedwater supplies.
- The use of chlorine gas is cheaper and cleaner.
- Chlorine supplies last approximately three months.
- Dosage rates and the resulting chlorine residual can be accurately controlled.

Disadvantages

- Boiling requires a reliable source of energy and is limited in terms of the volume able to be treated.
- The use of chlorine in gaseous form or in solution can cause safety hazards; all operating personnel should be made aware of these hazards and trained in their mitigation.
- Chlorine is reactive and interacts with certain chemicals present in the product water, depending on pH and water temperature; this results in the depletion of the chlorine concentration, leaving only residual amounts of chlorine for

disinfection (over-chlorination may result in the formation of chlorinated hydrocarbons, such as trihalomethanes, which are known to be carcinogenic).

- Chlorine will also oxidize ammonia, hydrogen sulfide, and metals present in the product water to their reduced states.
- Chlorine gas is heavier than air, and is extremely toxic and corrosive in moist atmospheres. Dry chlorine can be safely handled in steel containers and piping, but where moisture is present (as it is in most treatment plants), corrosion-resistant materials such as silver, glass, Teflon, and certain other plastics must be used - though not, for pressurized gas.
- Hypochlorite may cause damage to eyes and skin upon contact, and, because it is a powerful oxidant, may cause fires if it comes into contact with organic or other easily oxidizable substances.

Cultural Acceptability

Boiling is a widely accepted practice. Chlorination is a common practice in water treatment plants in urban areas, but is rarely used in rural areas.

Further Development of the Technology

Boiling and chlorination are very well known technologies used by most of the world's population for the routine and/or emergency disinfection of water supplies and wastewaters. Nevertheless, chlorination systems could be improved primarily in the area of safety both in the production of chlorine gas and the methods of handling and distributing the gas within the treatment plants.

Development of corrosion-resistant materials that are not affected by chlorine could increase the frequency of utilization of gas chlorination, which is a more efficient method of disinfection than hypochlorite. Hypochlorite production methods, using seawater and brackish water as source waters for the production of chlorine solutions, could also be improved, to reduce the cost and to make use of the by-products of this process.

FILTRATION SYSTEMS

Filtration systems are primarily used to purify water for domestic consumption. Several types of filtration systems have been used extensively in developing countries throughout the world, particularly in Latin America and the Caribbean. These include residential filters, slow and rapid sand filters, and dual media filters. Vertical flow pre-filters with gravel media tested in Guatemala and up-flow solids contact filters used in Brazil have potential for future use.

The design and application of different types of filters depend on the volume, flow rate, and quality of the inflowing water; the desired degree of water purification; and the use of the filtered water. The availabilities of filtering materials and skilled personnel are also factors to be considered in the selection of an appropriate filtration system.

Normally, the quality of the product water can be improved by mechanical straining through a porous material, such as sand or gravel. Depending on the size of the pores and the nature of the filter material, straining, or filtering, may remove a significant portion of the undesirable contents of the feedwater: suspended and colloidal matter, bacteria and other microorganisms, and, sometimes, certain chemicals.

The filter material may be any porous, chemically stable material, but sand (silica and garnet) is used most often. Sand is cheap, inert, durable, and widely available. It has been extensively tested and has been found to give excellent results. (Other materials have been used, some of which are described below; others, such as the reverse osmosis technologies described previously, are also a specialized form of filtration.)

Technical Description

Residential Filters

Residential filters are a common form of filtration. They can be either homemade or purchased commercially. The homemade filters usually consist of a sand- or gravel-filled pipe or tub, while the commercial systems usually have a stainless steel frame, with appropriate connections that make installation and operation relatively simple. Many commercial filters contain filtration media other than sand or gravel.

The basic form of residential filter, used in rural areas with no public water supply, is the tub filter. The tub filter consists of two tubs made of mud or clay, pottery or plastic, and joined together. The upper tub contains the filter medium (sand, gravel, coal, stone, etc.), into which the water to be treated is poured. It moves through the filter medium, through holes in the base of the upper tub, to the lower tub, where it is stored until used.

A faucet is usually installed in the lower tub for convenient access. Homemade filters, such as the tub filter, are usually constructed of locally available materials. For example, in El Salvador, they are constructed of a concrete pipe, approximately 0.5 m in diameter and 1 m in length, fitted with a perforated pipe, which is placed at the bottom of the filter in a 10 cm layer of gravel and connected to a pipe with a 3/4-inch internal diameter from which the filtered water is extracted.

The gravel is overlain by 60 cm of sand. Both the gravel and the sand are cleaned and dried in the sun, before use. In Mexico, residential filters are constructed of porous volcanic rock assembled in a wooden frame and protected by a screen. In the Dominican Republic, residential filters are installed at the point of discharge of storage cisterns, or at the point where water enters the houses.

The frame of these filters is usually made of stainless steel, with layers of sand, quartzitic gravel, anthracite, and activated carbon as the filtration media.

Slow Sand Filters

A slow sand filter consists of a watertight box, fitted with an underdrain, which supports the filtering material and distributes the flow evenly through the filter. Many different media have been used for the underdrain system. Bricks, stone, and even bamboo have been used for this purpose; bamboo, however, requires frequent replacement because it is organic and subject to decomposition.

The effective size of the sand used in slow sand filters is about 0.2 mm, and may range between 0.15 mm and 0.35 mm, with a coefficient of uniformity of between 1.5 and 3.0. In a mature bed, a layer of algae, plankton, and bacteria forms on the surface of the sand. The walls of the filter can be made of concrete or stone. Sloping walls, dug into the earth and supported or protected by chicken wire reinforcement and a sand or sand-bitumen coating, could be a cost-effective alternative to concrete.

Some Latin American countries, such as Ecuador and El Salvador, use concrete reinforced with a minimal amount of iron (ferrocement). Inlets and outlets should be provided with controllers to keep the raw water level and the filtration rate constant. Lateral pipes range from 2 to 8 in, while the bottom drains are normally between 10 and 30 in. Bottom drains consist of a system of manifold and lateral pipes is a diagram of a typical slow sand filter.

The successful performance of a slow sand filter depends mainly on the retention of inorganic suspended matter by the straining action of the sand. Filtration rates usually employed in developing countries range between 2.5 and 6.0 m^3/m^2/day. Higher rates may be used after a series of tests demonstrates that the effluents are of good quality.

The system should be designed for flexibility, and should consist of a number of separate units to enable maintenance to be performed without interruption of the water service. The suggested number of units for a given population size ranges from two units for a population of 2 000 up to six units for a population of 200 000.

Rapid Sand Filters

Rapid sand filters differ from slow sand filters in the size of the media employed. Media in rapid sand filters may range in size from 0.35 to 1.0 mm, with a coefficient of uniformity of 1.2 to 1.7. A typical size might be 0.5 mm, with an effective size of 1.3 to 1.7 mm.

This range of media size has demonstrated the ability to handle turbidities in the range of 5 to 10 NTU at rates of up to 4.88 $m^3/m^2/h$. Filtration rates for rapid filters may be as high as 100 to 300 $m^3/m^2/$day, or about 50 times the rate of a slow sand filter. The number of filters used for a specific plant ranges from 3 filters for a plant capacity of 50 l/s to 10 filters for a plant capacity of 1 500 l/s.

A typical rapid sand filter consists of an open watertight basin containing a layer of sand 60 to 80 cm thick, supported on a layer of gravel. The gravel, in turn, is supported by an underdrain system. In contrast to a slow sand filter, the sand is graded in a rapid rate filter configuration. The sand is regraded each time the filter is backwashed, with the finest sand at the top of the bed.

The underdrain system, in addition to performing the same functions served in the slow rate filter, serves to distribute the backwash water uniformly to the bed. The underdrain system may be made of perforated pipes, a pipe and strainer, vitrified tile blocks with orifices, porous plates, etc.

A clear well is usually located beneath the filters (or in a separate structure), to provide consistent output quantity. The minimum number of filter units in a system is two. The surface area of a unit is normally less than 150 m^2. The ratio of length to width is 1.25 to 1.35.

Dual- or Multi-Media Filters

Dual-media filtration uses two layers, a top one of anthracite and a bottom one of sand, to remove the residual biological floe contained in settled, secondary-treated wastewater effluents and residual chemical-biological floe after alum, iron, or lime precipitation in potable water treatment plants. It is also used for tertiary or independent physical-chemical waste treatment in the United States and other countries.

Gravity filters operate by using either the available head from the previous treatment unit or the head developed by pumping the feedwater to a flow cell above the filter cells. A filter unit consists of an open watertight basin; filter media; structures to support the media; distribution and collection devices for influent, effluent, and backwash water flows; supplemental cleaning devices; and the necessary controls to sequence water flows, levels, and backwashing.

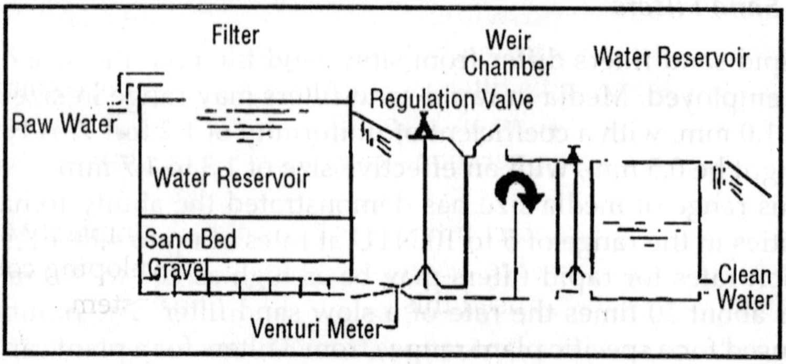

Fig. Slow Sand Filtration System

Upflow Solids Contact Filter

These units eliminate the need for separate flocculators and settling tanks, since they perform liquid-solid separation, filtration, and sludge removal in a single unit process. Coagulation and flocculation are performed in a granular medium (such as a layer of gravel under a sand bed).

The use of flocculent aids improves filtration results. This process should be restricted to raw waters of low turbidity (up to 50 JTU) and no more than 150 mg/l of suspended solids. It is widely used, especially in Brazil. These filters are designed for rates of filtration between 120 and 150 m^3/m^2/day.

Extent of Use

Both homemade and commercially purchased residential filters are commonly used in developing countries where the quality of water for domestic use is poor. El Salvador, Dominican Republic, and Mexico have promoted the use of these types of filters. In general, most Latin American countries use residential filters for water purification, particularly in rural areas.

Slow and rapid sand filters have been used in the rural community of La Pinera, El Salvador. In Ecuador, slow sand filters are used extensively for both surface and groundwaters. Filtration systems using vertical reactors with gravel beds have been tested as a means of pre-filtration in a water treatment plant in the municipalities of Cabañas and Zacapa, Guatemala.

Rapid sand filters are more complex to operate than then-slow sand filter counterparts, but they are widely used, especially in areas with high turbidity and where land requirements may be an important design consideration. Conventional rapid sand filtration plants are widely available and widely used in Latin America and other developing countries throughout the world.

Dual or multimedia filters are limited to developing countries that can inexpensively acquire anthracite. The higher skill level and energy requirements for the operation of these high rate systems may limit their application.

Upflow solids contact filters, because of their simplicity and low cost, could be an effective technology in many developing countries. Brazil has successfully used this type of filtration system.

OPERATION AND MAINTENANCE

The filter media of homemade residential filters must be periodically changed to maintain the filter's effectiveness. Most of the residential filters acquired commercially can be purchased with a maintenance contract, which will prolong their operational life.

A number of factors affect the operation and maintenance of slow sand filters. The initial resistance (loss of head) of a clean filter bed is about 6 cm. During filtration, impurities are deposited in and on the surface layer of the sand bed, and the loss of head increases. At a predetermined limit (the head loss is usually not allowed to exceed the depth of water over the sand, or about 1 m to 1.5 m), the filter is taken out of service and cleaned.

The period between cleaning is typically 20 to 60 days. The filter can be cleaned by either scraping off the surface layer of sand and replacing it with washed sand stored after previous cleanings (periodic re-sanding of the bed), or washing the sand in place with a washer that travels over the sand bed. If sand is readily available, the former method is favored; workers with wide, flat shovels do the scraping, removing 1 to 2 cm of the topmost material.

The amount of time this takes depends on the area of the filter bed, but it can usually be completed in one or two days. After washing, the sand is stored and replaced on the bed when, after successive cleanings, the thickness of the sand bed has been reduced to about 50 to 80 cm.

A sand and gravel filter needs to be replaced every two years or so. When using the method of washing in place, about 0.2% to 0.6% of the water filtered is required for washing purposes. The bacteriological layer, which is the most important layer in the filtration process, needs to be reactivated in the new filter. Reactivation usually lasts two months.

Rapid sand filtration plants are complicated to operate, requiring operator training in order for the plant to produce a product water of consistent quality and quantity. The filters require frequent backwashing to maintain satisfactory operating heads in the system (filter runs may vary from only a few hours to as many as 24 to 72 hours, depending on the suspended solids in the influent). Backwashing rates are typically 0.6 m^3/min or higher, for a period of several minutes.

In addition, the initial production following backwashing is channeled to waste for several minutes. Thus, the water backwashing uses can be as much as 10% to 15% of the total plant output. On the other hand, rapid sand filtration plants (including chemical treatment) can effectively treat higher solids loadings and produce higher outputs than slow sand filters. The land area requirements are significantly lower.

Dual-media filters, like rapid sand filters, are cleaned by hydraulic backwashing (upflow) with potable water. Thorough cleaning of the bed makes it advisable in the case of single medium filters, and mandatory in the case of dual- or mixed-media filters, to use auxiliary scour or so-called surface wash devices before or during the backwash cycle. In dual-media and mixed-media beds, such additional effort is needed to remove accumulated floe, which is stored throughout the bed depth to within a few inches of the bottom of the fine media. Backwashing is generally carried out every 24 to 72 hours.

The optimum rate of washwater application is a direct function of water temperature, as expansion of the bed varies inversely with the viscosity of the washwater. For example, a backwash rate of 18 gpm/ft^2 at 20°C equates to 15.7 gpm/ft^2 at 5°C, and to 20 gpm/ft^2 at 35°C. The time required for backwashing varies from 3 to 15 minutes. After the washing process, water should be discharged to waste until the turbidity drops to an acceptable value. Few data are available on the operation and maintenance of the vertical reactor pre-filters tested in Guatemala, which remain in the experimental stage.

Other operational considerations relating to the use of filtration technologies include the use of flocculent aids. Coagulants such as alum, ferrous sulfate, and lime may be added to aid in the flocculation and sedimentation of particulates.

The coagulant dosage is generally determined from jar tests, and the chemicals are almost always added with rapid mix systems. In the case of water treatment plants, flocculation is usually performed ahead of the settling process to improve the effectiveness of this process.

Maintenance considerations include the resolution of a number of problems which can interfere with the consistent operation of sand filters. These problems often are due to poor design or operation of the filtration systems. The problems most often encountered and their possible solutions are as follows:

- Surface clogging and cracking: This problem, caused by an overload of solids at the thin filter layer in sand filters, can be alleviated by using dual or multiple media, which allows deeper penetration of solids into the bed, and, generally, longer run times.
- Gravel displacement or mounding: This problem can be alleviated by placing a 76 mm layer of coarse garnet between the gravel supporting the media and the fine bed material.
- Mudball" formation: This problem can be reduced by increasing the backwash flow rate (e.g., up to 20 gpm/ft^2), and by providing for auxiliary water or air scouring of the washed surface.
- Sand leakage: This problem may be alleviated by adding the garnet layer.
- Accumulation of air bubbles in the bed: This problem, which causes a significantly increased resistance to flow through the filter, can be minimized by maintaining adequate water depths in the clear well and filters; frequent backwashing may help.

LEVEL OF INVOLVEMENT

In many developing countries, filtration methods are introduced and promoted by both governmental agencies and NGOs, with the full participation of the community. This is the case in El Salvador, where the Centro Salvadoreño de Tecnología Apropiada (CESTA) builds and installs residential filters for rural communities at a minimum cost. In Dominican Republic, the private sector, particularly the companies which manufacture filtration systems, promotes the technology. In Ecuador, NGOs like Plan Internacional and CARE actively participate in the implementation of these technologies in order to reduce the use of contaminated water. In Brazil, the government and the private sector are actively involved in the development and implementation of filtration systems.

Costs

Homemade residential filters were constructed in El Salvador at a cost of $23. Operation and maintenance costs are about $6/year. The cost of residential filters manufactured and commercially distributed in Dominican Republic varies with the flow capacity of the filter.

It ranges from $382 for 1 gpm to $588 for 6 gpm; this price includes installation and maintenance. Commercially manufactured tub filters are sold in Dominican Republic hardware stores at a price ranging from $26 to $45. The cost of quarry filters used in Mexico was $50, with little or no operation and maintenance cost shows the construction cost of an upflow solids contact filter a function of the filtration area.

The unit filtering cost of slow and *rapid* sand filters in Ecuador ranges between $\$0.13/m^3$ and $\$0.20/m^3$. Slow sand filters were constructed in Ecuador at a cost of $132.30 with an estimated operation and maintenance cost of 25% of the construction cost. Table shows estimated per capita costs of construction and of operation and maintenance for slow and rapid sand filters.

EFFECTIVENESS OF THE TECHNOLOGY

Homemade residential filters can adequately reduce the level of contaminants in water, but, because quality control tests are usually not performed on the product water, there is a risk of some contamination remaining after filtration.

For example, quarry filters used in Mexico reduce bacteriological contaminants by up to 90%. However, quarry filters must be covered and protected with a screen, and a faucet at the outlet is recommended. This filter needs to be cleaned every 3 to 4 months, depending on the quality of the water treated.

Commercially available residential filters are usually more effective at producing a good quality product water since quality control is performed during the manufacturing process and a level of efficiency is initially guaranteed.

Quality product water can be further ensured through the regular inspections performed by technicians from the supplier in the case of systems sold with a service contract.

Slow sand filters are very effective in removing solids and turbidity when the raw water has low turbidity and colour (turbidity up to 50 NTU and colour up to 30 Pt units). Taste and odor are also improved. However, if the raw water quality is poor, filtration is often less effective. In such situations, roughing filters, or pre-filters, are often used before the feedwater enters the slow sand filters.

The slow sand filters are very effective in removing bacteria; in general, their effectiveness in removing bacteriological contaminants ranges between 80% and 99%, depending on the initial level of contaminants and the number and design of the filtration units. In many regions of Ecuador, the effectiveness is close to 100%.

In El Salvador, they are estimated to remove 84% and 99% of total and fecal coliform bacteria, respectively. Reductions in the levels of iron, manganese, and nitrate concentrations and turbidity are also observed. Chemicals are typically not used. The flow rates for slow sand filters are many times slower than for rapid sand and roughing filters, and the operating filter bed is not stratified. Multimedia filters are usually more effective, since the filtration media combine the filtration properties of several materials. In the system of vertical flow pre-filters used in Guatemala, turbidity reduction ranged from 23% to 45%, and colour reduction between 34% and 56%.

Suitability

Filtration technologies are suitable for use throughout the region. Homemade residential filters are better suited to rural areas where the equipment, skills, and infrastructure necessary to provide piped domestic water supplies are lacking. The other, more complex filtration systems are best used at water treatment plants and are generally located in urban areas.

Advantages

- Filtration systems have a low construction cost, especially when built using manual labour.
- These systems are simple to design, install, operate, and maintain, which makes them ideal for use in areas where skilled personnel are few.
- No chemicals are required, although flocculent aids are sometimes used in conjunction with large-scale filtration systems; supplies of sand can usually be found locally.
- Power is not required.
- Large quantities of washwater are not required.
- Use of filtration to pretreat water and wastewaters results in fewer sludge disposal problems because fewer contaminants are left to be removed during the treatment process.
- Residential filters provide adequate treatment of water for average-sized households, particularly in rural areas.
- Filters are environmentally friendly.

Disadvantages

- In some areas, there is a lack of locally available filtration media.

- There may be a lack of skilled personnel to operate the more sophisticated filtration systems, particularly in rural areas.
- Use of filtration alone is recommended only for source waters with low levels of contamination.
- Pretreatment may be required for many applications.
- Provision must be made for washing and storing used sand from sand filters, either permanently or temporarily, and for moving sand from the filters to the wash site and from the wash site to the storage site and back, as needed.
- If sand from slow sand filters is to be washed, a separate backwash facility and washwater supply may be required; treated water must often be used for washing, which could reduce the available supply of treated water, especially in water-poor areas.
- Precise operational control of the rate of head loss is required to prevent air bubbles from entering and binding the system; this type of interference is a potential problem in all types of filters.
- There may be a lack of quality control of the product water in rural areas.

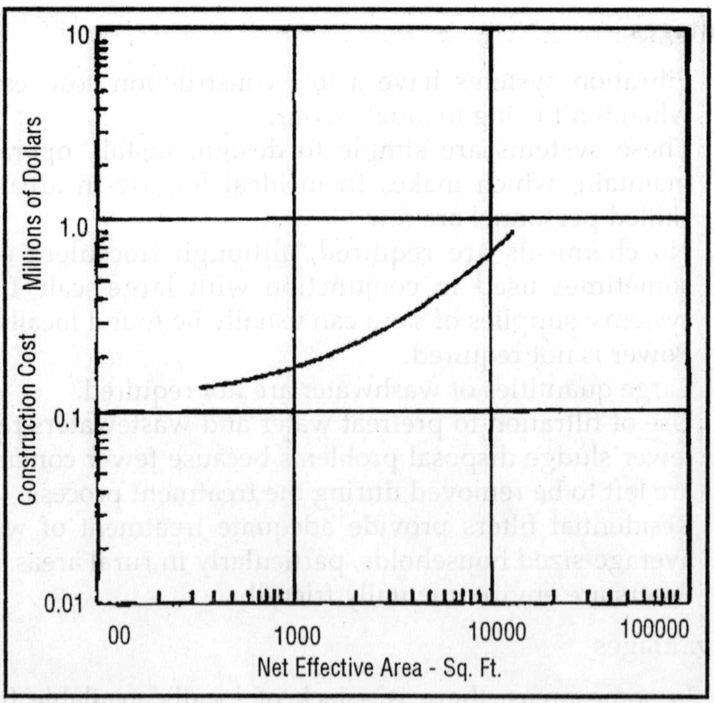

Fig. Construction Cost of Upflow Solids Contact Filter.

- To obtain good results from slow sand filters, the raw feedwater must not generally have a suspended solids content of less than 50 mg/1.

Cultural Acceptability

Filtration is a well-accepted technology when applied in the treatment of industrial and public water supplies. It has limited acceptance in other applications, and at the household level in rural areas.

FURTHER DEVELOPMENT OF THE TECHNOLOGY

Additional research is needed to develop more efficient filtration media that can remove both bacteriological and chemical contaminants. Education is needed, particularly in the rural areas, to encourage the use of homemade filtration systems and disinfection of household water supplies.

Chapter 10

Biological Significance of Prostaglandins

Eicosanoid is an umbrella term for all of the biologically active metabolites of arachidonic and two other C20 polyunsaturated fatty acids (PUFAs). Insect scientists are most familiar with the prostaglandins, which make up one of several groups of eicosanoids. PGs are responsible for release of egg-laying behaviour in a few insect species, and PGs and certain other eicosanoids appear to mediate hemocytic immune responses to bacterial infections in insects. PGs were recently implicated in regulation of Malpighian tubule physiology.

The significance of our increasing knowledge of eicosanoids is that they represent a tier of regulatory physiology in invertebrates that has largely gone unrecognized hitherto.

On that basis, we gain from our studies of eicosanoids a more complete understanding of the complexity of regulation of invertebrate physiology.

Because most of our background knowledge of eicosanoids arises from work on mammalian physiology and pathophysiology, we also gain a broader appreciation of comparative physiology.

As an index of the growing awareness of the importance of eicosanoids in insects and other invertebrates, this and the following two chapters in this volume are devoted to the subject.

Brady reviewed the very early findings on PGs in insects; Bundy discussed non-mammalian sources of PGs; Stanley- Samuelson and Loher reviewed the details of PGs in insect reproduction, which remains contemporary. In the last several years. Two reviews present a broader focus with respect to fatty acid metabolism and eicosanoids in insects. A comprehensive review of eicosanoids in insects is forthcoming, which will cover fatty acid nutritional requirements and their relationship to eicosanoids, a detailed history of the field, and a discussion of the ecological significance of eicosanoids.

EICOSANOID BIOSYNTHESIS IN MAMMALS

Eicosanoid biosynthesis is grounded in the nomenclature and metabolism of fatty acids. The structures, names and standard notation for individual fatty acids are summarized in most insects are thought to biosynthesize C20 PUFAs from their C18 counterparts, and these pathways are outlined in Background information on fatty acid nutritional requirements, the occurrence of C20 PUFAs, and on PUFA metabolism in insects is available.

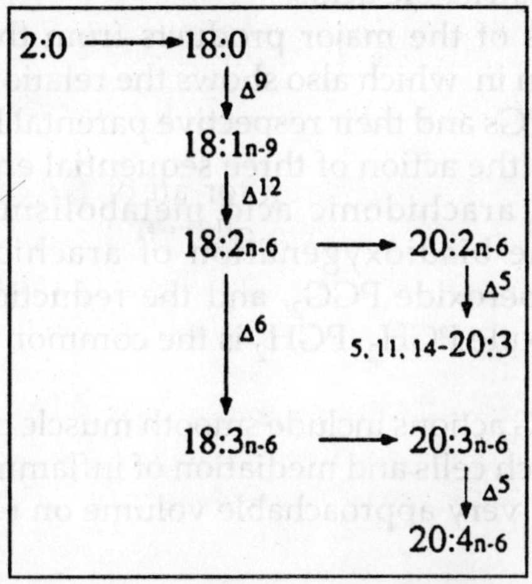

Fig. Elongation/desaturation Pathways for Biosynthesis of PUFAs in Insects

More detailed treatment than presented herein on the biosynthesis and structures of eicosanoids, and on their biological action in mammals, can be found in the chapters on phospholipases and eicosanoids in Vance and Vance, and in the reviews by Samuelsson*et al.* Samuelsson, Moore, Needleman*et al.*, Spector*et al.*, Pace-Asciak and Asotra, McGiff, and Holtzman.

Cellular and pharmaceutical regulation of PLA_2 activity is an independent area of intense research effort. Upon stimulation by various agonists, many mammalian cells hydrolyze PUFAs, by action of PLA_2, from the *sn*-2 position of membrane phospholipids (PLs). PLA_2 is regarded as a crucial regulating step in cellular PL metabolism, protection of membranes from peroxidation damage and in eicosanoid biosynthesis.

Three C20 components — dihomo-*gamma*-linolenic (20:3n-6), arachidonic (20:4n-6) and eicosapentaenoic (20:5n-3) acids — are potential substrates for eicosanoid biosynthesis. The best known substrate is arachidonic acid, which may be converted to eicosanoids by three major pathways:

- The cyclooxygenase pathways, which lead to PGs and thromboxanes (TXs).
- The lipoxygenase pathways, leading to hydroperoxy- and hydroxypolyenoic fatty acids (HPETEs and HETEs), which are themselves biologically active as well as potential substrates for further metabolism to leukotrienes (LTs) and other biologically active products;
- The cytochrome P-450 epoxygenases, which produce epoxyeicosatrienoic acids.

The structures of the major products from the cyclooxygenase pathway are shown in which also shows the relationship between the 1-, 2- and 3-series PGs and their respective parental PUFAs. Any given PG is produced by the action of three sequential enzymatic steps.

Focussing on arachidonic acid metabolism, PGH_2 synthase catalyzes both the bisdioxygenation of arachidonic acid to the hydroperoxyendoperoxide PGG_2, and the reduction of PGG_2 to the hydroxyendoperoxide PGH_2. PGH_2 is the common precursor to all of the 2-series PGs.

Examples of PG actions include smooth muscle contraction, release of acid from stomach cells and mediation of inflammation physiology. Moore produced a very approachable volume on relevance of PGs in mammals.

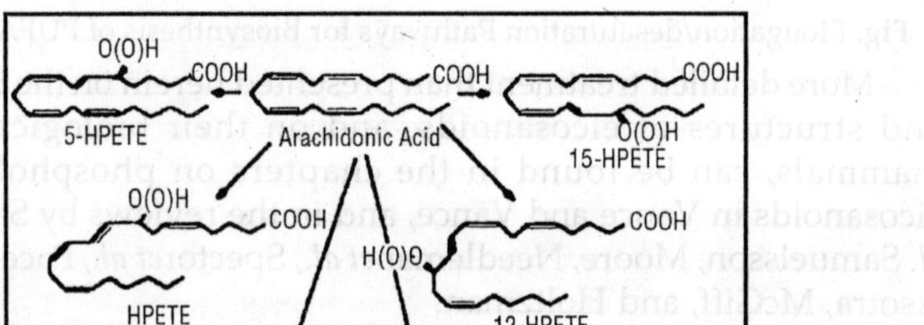

Fig. The structures of HPETEs and HETEs.

The structures of arachidonic acid metabolites of the lipoxygenase pathways are shown in Similar products are formed from 20:3n-6 and 20:5n-3, but the individual structures are not shown. As indicated, a separate lipoxygenase activity, designated by the position of the functional group, is responsible for synthesis of each product. Lipoxygenase pathways first produce HPETEs, which are themselves biologically active, as well as further metabolized to HETEs. Di- and trihydroxy fatty acids can be formed by actions of more that one lipoxygenase. Some HETEs are metabolized into their respective keto analogues, KETES, at least one of which is biologically active in invertebrates. Aside from recently implicating lipoxygenase products in insect immunity, we know very little about lipoxygenase systems in insects. Lipoxygenase activity has been reported in a fruitfly and a firebrat.

The 5-lipoxygenase pathway is of special interest because 5-HPETE can be dehydrated to form LTA $_4$, the root compound for synthesis of all LTs. These lipoxygenase reactio ns importantly in mammalian host defence systems such as the various leukocytes, macrophages, monocytes, and cells of lung and spleen.

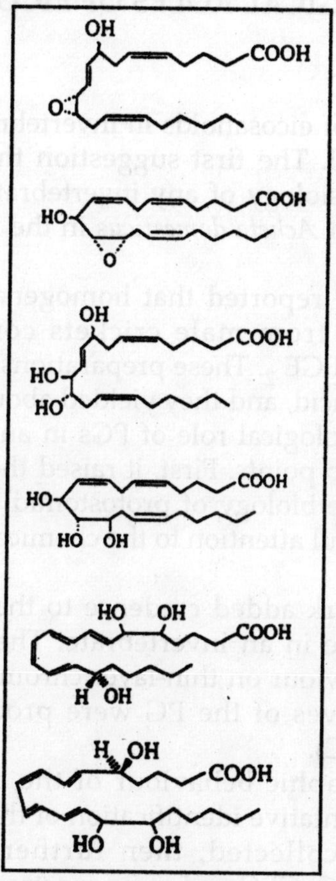

Fig. Structures of Lipoxins, Hepoxilins and Trioxilins

For examples, some HETEs are active in inducing the chemokinesis and chemotaxis associated with defence cell migration, and the slow-reacting substance of anaphylaxis is a mixture of LTs. Other lipoxygenase products, which are biologically active, include lipoxins, hepoxilins, and trioxilins. The hepoxilins, for example, are formed in pancreatic islets, where they act to release insulin. They may be similarly active in hormone release in invertebrates.

Eicosanoids are involved in basic physiological processes, mainly at the cellular level, in mammals. Moreover, they appear to be especially important in various pathophysiological responses such as inflammation, blood-clotting, hypersensitivity reactions and tumor growth. The information presented here, which may be taken as the mammalian model, is meant to convey an idea of the biosynthesis and structures of eicosanoids.

PHYSIOLOGICAL ROLES OF EICOSANOIDS

REPRODUCTION

The significance of eicosanoids in invertebrate reproduction has been treated in detail. The first suggestion that eicosanoids were important in the physiology of any invertebrate animal came from research on the cricket *Acheta domesticus* in the laboratory of the late U.E. Brady.

Destephano *et al.* reported that homogenates of frozen whole reproductive tracts from male crickets converted exogenous arachidonic acid into PGE_2. These preparations were incubated with 600 ìg of arachidonic acid, and they yielded about 16 ìg of PGE_2. This first work on a physiological role of PGs in an invertebrate holds a special interest on two points. First, it raised the possibility that PGs were important in the biology of protostomid animals; and second, the authors paid careful attention to the chemical identification of the PG structures.

The chemical work added credence to the idea that PGs were physiologically active in an invertebrate. The PG was tentatively identified by its behaviour on thin-layer chromatography (TLC) and appropriate derivatives of the PG were processed by gas-liquid chromatography (GLC).

The chromatographic behaviour of the derivatives was also consistent with the tentative identification of the PG. The GLC peaks were individually collected, then further analyzed by mass spectrometry. The TLC, GLC and mass spectrometric data permitted the authors to conclude that the male reproductive tracts from the cricket *A. domesticus* biosynthesized PGE_2. The authors speculated that the PG might be involved in sperm transport within female insects.

Information on a physiological action of PGs in invertebrates emerged when Destephano and Brady demonstrated that PGE_2 released egglaying behaviour in newly-mated females of *A. domesticus*. Among other constituents of seminal fluids, PG biosynthetic activity is transferred from males to females in spermatophores. The activity can be detected in spermathecae of mated females, and it is responsible for conversion of arachidonic acid into PGE_2. Although the mechanism remains unclear, the PGE_2 formed within the spermathecae ultimately releases the full program of egg-laying behaviour.

PGE_2 also releases egg-laying behaviour in newly mated females of the Australian field cricket *T. commodus*. The enzyme-transfer mechanism of releasing egg-laying behaviour is similar in both cricket species. Stanley-Samuelson and Loher developed a more detailed picture of the mechanism. They noted in quantitative GLC studies that spermathecae from mated crickets contained more arachidonic acid than spermathecae from virgins.

The arachidonic acid could arise from increased biosynthesis from shorter chain-length components, or it could have been transferred from males along with other contents of seminal fluids. Upon analysis of spermatophore lipids by quantitative GLC, Stanley-Samuelson and Loher concluded that spermatophores provided the increased spermathecal arachidonic acid. All of the arachidonic acid in spermatophores was associated with PLs. Within PL species, arachidonic acid comprised about 25% of PC fatty acids and about 4% of PE fatty acids. Once within spermathecae, the arachidonic acid could be hydrolyzed from the PC and then converted into PGs.

The transfer of arachidonic acid during mating was confirmed using radioisotope methodologies. Radioactive arachidonic acid (20 ìCi) was injected into the hemocoels of adult males within 8 hr of emergence. These males began producing spermatophores on the usual schedule for untreated individuals, 3 to 5 days after emergence.

Spermatophores were collected, total lipids extracted, and the radioactivity in the total lipid extracts determined by liquid scintillation counting. Spermatophores were collected on an irregular basis for the next 49 days, and all contained radioactivity. Radioactivity in some spermatophores was characterized on radio-HPLC that was optimized to separate fatty acid methyl esters. These analyses showed a single peak of radioactivity that corresponded with authentic arachidonic acid.

Midway through this experiment, the radioactive males were individually placed into mating arenas, and allowed to mate with untreated females. The spermathecae were isolated from females at either 1 hr or 24 hrs after mating. The spermathecae isolated 1 hr after mating contained about 700 cpm of radioactivity, which, based on our studies, would be expected if all the radioactivity normally found in a spermatophore were transferred into a spermatheca.

The spermathecae isolated 24 hours after mating contained only background radioactivity, but radioactivity was recovered from hemolymph and ovaries of these females. About half of the hemolymph radioactivity co-chromatographed on silicic acid column chromatography with authentic PGs and half with authentic arachidonic acid.

Eicosanoids are not sufficiently resolved on silicic acid chromatography to claim rigorous identification of these products. However, these results indicate that radioactive arachidonic acid is transferred to spermathecae, partly metabolized to PGs, and subsequently moved from spermathecae into hemolymph circulation.

Loher and Edson showed that testectomized males remain able to produce spermatophores and mate with females. However, these matings do not release egg-laying behaviour. Spermatophores from testectomized males do not contain PG biosynthetic activity, whereas spermatophores from sham-operated males contain the usual level of PG biosynthetic activity.

These results indicate that formation of spermatophores with this activity depends on the testes. Stanley-Samuelson*et al.* investigated the possibility that testes are required for transfer of arachidonic acid. Males were testectomized shortly after adult emergence, then injected with 10 ìCi of radioactive arachidonic acid. The treated males produced spermatophores which had amounts of radioactivity similar to the amounts found in spermatophores from intact males.

These experiments indicate that the testes are probably the source of PG biosynthetic activity and accessory glands may be the source of the arachidonic acid in spermatophores. Because testes are not required to produce radioactive spermatophores, we speculated that accessory glands package arachidonic acid during spermatophore biosynthesis. A direct test of this idea is not feasible because accessory glands are involved in producing spermatophores.

PGs may release egg-laying behaviour in females of the silkmoth *Bombyx mori*. Yamata Setty and Ramaiah explored this by injecting PGs into the abdomens of virgin females the day after adult emergence, then observing the number of eggs that were deposited during the following 24 hours. Saline-injected controls laid 19 eggs/female (mean, n=12).

Virgins that were treated with 1, 10 and 100 ìg of PGE2 laid, respectively, 32, 42 and 63 eggs/individual (mean, n> 10). These are statistically significant increases in oviposition, compared to controls. There is a question on the biological significance of these numbers because normal matings result in deposition of about 500 eggs/female.

The highest PGE2 dose resulted in about 12% of the number of eggs expected from a mated female. There is also a question of specificity because 100 ìg doses of PGE1 and PGF2á also stimulated release of 46 and 47 eggs/female (mean, n> 10). These results are similar to the PGE2 effects.

Experiments with PG biosynthesis inhibitors support the idea that PGs release egg-laying behaviour in newly mated female silkmoths. Males and/or females were injected with 10ìg of aspirin or indomethacin, then allowed to mate. The numbers of eggs laid during the following 24 hours were then counted. Matings between untreated or control groups produced over 500 eggs/individual.

However, treating males and females with inhibitors resulted in substantial reduction in oviposition; by 88% with indomethacin and by 66% with aspirin. Matings between treated females and normal or control males did not reduce egg-laying. Reciprocal matings, between treated males and control females, reduced oviposition to the levels observed when both sexes were treated.

The authors also observed that injecting PGE_2 into females that had been mated with treated males, reversed the effects of the indomethacin treatments. The PGE_2 increased, but did not completely restore, normal egg-laying. The authors suggested that PGs, rather than PG biosynthetic activity, are transferred from male to female, resulting in increased oviposition behaviour.

These data indicate that PGs are involved in releasing egg-laying behaviour in silk moths. Treating virgin females with PGE_2 resulted in very little increase in egg-laying, while treating males with aspirin or indomethacin substantially reduced the normal level of egg-laying that follows mating.

Aspirin and indomethacin inhibit cyclooxygenase activities in mammals, and possibly in insects. Inhibition at this point in the PG biosynthetic pathways would reduce synthesis of all PGs. It may be that PGs other than PGE and PGF are more active in releasing egg-laying behaviour. Alternatively, the quantitative relationship between PGE and PGF may be important in releasing egg-laying.

The emergent picture is that PGs mediate egg-laying in a few insect species, certainly not most of them. Moreover, the specific mechanisms of releasing egg-laying behaviour differ among insect species.

One point, as detailed below, is that PGs occur in the reproductive tracts of all insect species that have been so investigated, including many in which PGs do not mediate egg-laying. And secondly, PGs are involved in basic aspects of reproduction in many other invertebrates.

PGs or PG-biosynthetic activity are sexually transferred to females of a number of insect species including the cabbage looper, *Trichoplusia* ni, the locust Locusta migratoria, the housefly *Musca domestica* and the hematophagus bug *Triatoma infestans*. Increased oviposition behaviour does not follow the transfer of PGs.

I will review the details of this work, then place these observations into the broader context of the roles of PGs in reproductive biology. The central point that PGs may be involved in many aspects of reproduction will be underscored by drawing upon literature from other invertebrate animals.

Compared to the reproductive tracts of virgins, the reproductive tissues of mated females of the cabbage looper *T. ni* contain increased quantities of PGE_2 and $PGF_{2á}$. Based on the results from another lepidopteran, the silkmoth, it might be thought that PGs, rather than PG-biosynthetic activity, are sexually transferred in *T. ni*. Studies on PG biosynthetic activity were not reported, but it seems that the PG increases are due to transfer of PG biosynthetic activity.

First, the reproductive tracts from virgin males and females had about the same amount of PGE_2; tissues from mated females contained almost three times more. Second, compared to virgins, tissues from mated females contained twice as much $PGF_{2á}$, but none was detected in reproductive tissues from virgin males. Although other mechanisms are possible, these increases are conservatively explained by transfer of PG biosynthetic enzymes during mating. If this is so, then PGs are transferred differently in silkmoths and the cabbage looper moth. Treating virgin females with 100 ìg injections of PGE_1, PGE_2 and $PGF_{2á}$ did not increase egg-laying.

PG-biosynthetic activity is sexually transferred to females of *L. migratoria*. Using the assay of Loher*et al.*, Lange showed that PG-biosynthetic activity was present in spermatophores, opalescent glands and seminal vesicles of adult males; activity was not detected in accessory glands other than the opalescent glands, nor in testes. Low levels of activity were detected in spermathecae of virgins, which increased by an order of magnitude after mating. Again, PG treatments did not stimulate egg-laying in sexually mature virgin locusts.

Male and female houseflies feature low proportions of arachidonic acid in PLs. Houseflies also sexually transfer eicosanoids, but as precursor PUFAs, rather than biosynthetic enzymes or PGs *per se*. When female houseflies were allowed to mate with males that had been prelabeled with radioactive arachidonic acid the amounts of radioactivity recovered from females 4 hours after mating were consistently 3 to 5 times more than the amounts of radioactivity in male reproductive tracts.

The authors suggested that internal arachidonic acid is mobilized during the males lengthy mating period, transferred to male reproductive tract, and subsequently transferred to the female. Most of the transferred radioactivity was incorporated into lipid fractions within females, and small proportions were converted into several PGs, including PGE_2, $PGF_{2á}$ and PGA_2/B_2. The possibility that these PGs may release egg-laying behaviour remains unexplored.

Mating-stimulated increases in PGs have been observed in the reproductive tracts of female insects, as just described for representatives of Orthoptera, Lepidoptera and Diptera. suppose that such increases occur in most insect species. So far, it appears that eicosanoids may not increase in females of the German cockroach *Blattella germanica.*

Substantial quantities of PGs occur in virgin females of this species and there is no increase upon mating. However, these analyses were carried out on whole animals, which may obscure small changes in PG titres within the female's reproductive tract. As we have just seen, one biological effect of these increased PG titres is the release of egg-laying behaviour in some species.

The following examples from lower vertebrates and invertebrates will help develop a broader view of the significance of PGs and other eicosanoids in invertebrate reproduction. The point to be offered is that it can be reasonably speculated that PGs play important, and far ranging, roles in the reproductive biology of all animals. Viewed in this context, the significance of eicosanoids in insect reproduction can be expected to extend far beyond the effects of PGs on the egg-laying behaviour observed in a limited number of insect species.

PGs are also active in reproductive biology of fishes and reptiles. Ovulated female goldfish release F-type PGs into the water. These PGs constitute the goldfish postovulatory pheromone, and they stimulate male spawning behaviour. Other arachidonic acid metabolites appear to mediate release of gonadotropin in goldfish. Similarly, arginine vasotocin induces PG biosynthesis in reproductive tracts of the viviparous lizard *Sceloporus jarrovi.*

The PGs are thought to contribute to stimulation of oviductal contractions associated with parturition in reptiles. $PGF_{2á}$ titres approximately doubled in the blood of female garter snakes by about

15 minutes following mating. The increased PG titres are thought to have a functional role in inducing post-mating declines in sexual behaviour.

These examples are not meant to form an exhaustive review. They are lifted from the literature to illustrate the point that eicosanoids are involved in diverse aspects of vertebrate reproduction. Taken with the information on invertebrates, it is reasonable to propose that PGs are fundamentally important in reproduction in all animals.

IMMUNITY

Eicosanoids play central roles in the inflammatory and immune responses of mammals. Among other important roles in immunity, eicosanoids potentiate inflammatory reactions such as vasodilation, which induces edema, and biosynthesis of complement proteins. Many eicosanoids, including cyclooxygenase and lipoxygenase products mediate hypersensitivity reactions in mammals and are involved in natural killer cell anti-cancer actions.

Besides their roles as mediators of immune responses, PGs act as physiological modulators of immune functions in mammals. The importance of eicosanoids in mammalian immunity responses in underscored by the recognition of their growing importance in clinical settings.

The roles of eicosanoids in immunity of non-mammalian vertebrates are coming to light. Given the importance of these molecules in the immunity of most vertebrate classes, mediation of immune responses may be another area of animal biology in which eicosanoids are fundamentally and intricately involved.

Based on such reasoning, Stanley-Samuelson*et al.* put forth the view that eicosanoids are mediators of hemocytic immunity in insects and, by extension, in other invertebrates. To place this work in a broader perspective, the volumes edited by Gupta and Brehelin and the reviews by Ratcliffe, Dunn, Boman and Hultmark, Lackie and Karp provide the following sketch of invertebrate immunology.

Invertebrate immunity is an amalgam of several distinct physiological systems that provide protection from potential pathogens. Bacterial infections induce insects to synthesize a number of anti-bacterial proteins.

These include lysozymes, enzymes that directly attack bacteria by hydrolyzing the peptidoglycan cell walls of bacteria and a number of antibacterial proteins that function by disrupting bacterial cell membranes. These include cecropins (4 kDa), attacins (21 to 23 kDa), diptericins (8 kDa) and insect defensins (4 kDa). One of the proteins induced in the silk moth *Hyalophora cecropia* was designated P4. This protein is temporally one of the first proteins that bind to bacteria. Recently Sun*et al.* recognized that this protein belongs to the immunoglobulin superfamily of proteins. They re-named the protein *hemolin,* and suggested that it takes part in formation of a protein complex that initiates the immune response. Insects also produce agglutinating proteins, known as lectins, which may serve to agglutinate invading micro-organisms.

Anti-bacterial proteins are regarded as part of the immune response to bacterial infections. Recently, however, Russell and Dunn showed that lysozyme is released into the pupal midgut lumen of *M. sexta* in the absence of bacterial infection. The authors suggested that the high midgut lysozyme activity confers prophylactic protection from bacterial infections during metamorphosis. Expression of the midgut lysozyme gene appears to be under developmental control.

In addition to the humoral immunity conferred by various antibacterial proteins, insects elaborate several forms of cellular, or hemocytic, immunity. There is a very large literature on the details of hemocyte classifications and nomenclatures, which may be approached in the volumes edited by Gupta and in the review by Stanley-Samuelson.

When insects, and other invertebrates, are challenged with foreign materials, hemocytes respond in any of three ways: phagocytosis, encapsulation and nodule formation. Beside these direct interactions with invaders, hemocytes indirectly contribute to immune responses. These include hemolymph coagulation at wound sites, release of the prophenoloxidase activating system and detoxification reactions.

Hemocytic actions are the immediate responses to bacterial infections. Anti-bacterial proteins are not detectable in hemolymph until 6 to 12 hours after infection. For example, when living bacteria were injected into larvae of the tobacco hornworm *M. sexta,* the bacteria were rapidly removed from the circulating hemolymph. The clearance was accomplished through cellular actions.

Nodule formation was the prominent cellular action during the first two hours, followed by phagocytosis during the next 6 hours. These differences in response time courses can be used to experimentally dissect humoral and hemocytic responses to bacterial infections.

Shows that each of the major eicosanoid biosynthetic pathways may be probed by application of appropriate pharmaceutical compounds. These inhibitors were used to test the idea that eicosanoids are involved in mediation of cellular actions that clear bacterial infections from the hemolymph circulation of *M. sexta* larvae. emphasize that use of such inhibitors is based on the assumption that they exert similar actions in mammals and invertebrates.

Although we argued that this was probably so for the inhibitors we employed, the pharmacology of such compounds in any invertebrate has not been elucidated. In preliminary experiments, test larvae were treated with 50 ìg of the PLA $_2$ inhibitor dexamethasone and control larvae were treated with ethanol.

Then both groups were infected by intrahemocoelic injections of a red-pigmented strain of the insect pathogen *Serratia marcescens* (ca. 5 X 10^5 bacterial cells/larva). Hemolymph samples were withdrawn, diluted and streaked on agar plates at 15, 30, 45 and 60 minutes post-infection.

The 50 ìg of dexamethasone that were injected in these initial experiments are high, and potentially misleading dosages. The effects of far lower doses, from 1.4 X 10^{-5} to 0.14 ìg/larva, were subsequently ›xamined. The experiments were similar to those just described, with the exception that hemolymph was sampled once, 60 minutes post-infection.

Inhibition of eicosanoid biosynthesis appeared to compromise the ability of larvae to clear bacteria from their circulating hemolymph. It may, on these grounds, be expected that similar inhibition of eicosanoid biosynthesis would also result in increased larval mortality. This idea was tested by treating groups of test larvae with the dexamethasone dosages just described, then injecting them with bacteria.

Eicosanoid biosynthesis depends upon availability of free substrate, which for the most part is hydrolyzed from membrane PLs by action of PLA $_2$. Dexamethasone is thought to indirectly inhibit eicosanoid biosynthesis by inhibiting PLA $_2$ activity. Dexamethasone compromises the cellular immune responses to bacterial infections.The dexamethasone effect was reversed by treating the larvae with arachidonic acid. In these experiments, larvae were treated with the highest dose of dexamethasone (0.14 ìg/larva), then treated with 50.0 ìg of arachidonic acid. When these larvae were infected with bacteria, those treated with arachidonic acid had significantly reduced larval mortality. The ability to reverse the effects of the inhibitors adds considerable support to the idea that eicosanoids are crucial mediators of insect cellular immunity.

Dexamethasone inhibits total eicosanoid biosynthesis through its action on PLA $_2$ activity. The effects of dexamethasone do not allow us to draw conclusions on the relative importance of individual eicosanoid- biosynthetic pathways in cellular immunity. We probed the cyclooxygenase and lipoxygenase pathways by treating separate groups of test insects with indomethacin (cyclooxygenase inhibitor) and esculetin (lipoxygenase inhibitor).

The results of these experiments showed that inhibition of each pathway impaired the hornworms' ability to clear bacterial infections from their hemolymph, however, the degree of impairment did not match the dexamethasone results. We interpreted these results to suggest that products of both major pathways were involved in the organismal response to bacterial infections. Indeed, when larvae were treated with a combination of indomethacin and esculetin, the results were identical to those of experiments with dexamethasone.

We return, again, to the point that the occurrence and biological significance of arachidonic acid in insects is not yet widely appreciated. When eicosanoid mediation of an important physiological action is proposed, it is necessary to address basic issues of eicosanoid-biosynthesis and eicosanoid-precursor fatty acids. Preliminary evidence on these points was presented.

Ogg*et al.* studied the fatty acid compositions of lipids prepared from the hemocytes of *M. sexta* larvae and showed that the fatty acid compositions of hemocytes are different from the serum lipids and from the insects' culture medium. Small quantities of eicosarnoid precursor fatty acids, 20:3n-6, 20:4n-6 and 20:5n-3 were detected in total lipids, phospholipids and phospholipid fractions. Although the amounts were quite low, they were consonant with the emergent picture of fatty acid compositions of terrestrial insects.

The hemocytes were also able to incorporate exogenous radioactive arachidonic acid into cellular phosphohpids. The radioactive arachidonic acid was selectively incorporated into individual phospholipid species; most into PC, less into PE and still less into PS/PI (these two fractions co-chromatograph on one-dimentional TLC).

These differences in uptake are at least partly due to the differential amounts of individual phospholipid species within cells. In lepidopterans so studied, the choline and ethanolamine fractions are the predominant phospholipids, and PC occurs in higher proportions than does PE. Similar differences in incorporation of eicosanoid-precursor polyunsaturated fatty acids have been recorded in mammalian cell systems, and these differences are taken to indicate that the proportions of eicosanoid precursors in phospholipid species are individually regulated within cells.

THERMOBIOLOGY

Many mammals express fevers as one line of defensive response to infections. Fever also occurs in non-mammalian vertebrates, although the increased body temperatures may be mediated by behavioral, rather than physiological, mechanisms. Behavioral fever has been observed in frogs, a lizard and several fishes.

There are no data on the role of eicosanoids in febrile responses in insects. However, some invertebrate animals, including insects, express behavioral fevers. Here information is presented from other invertebrates, which allows the speculation that eicosanoids mediate fevers in insects. Following bacterial infection, the freshwater crayfish *Cambarus bartoni* produced a 2°C fever by moving into a zone of warmer water.

This response may be mediated by endogenous PG biosynthesis because increasing doses of PGE_1 induced similar fevers in uninfected individuals. Three marine arthropods, the American lobster *Homarus americanus*, the pink shrimp *Penaeus duorarum*, and the horseshoe crab *Limulus polyphemus*, similarly increased their body temperatures after injections of 100 ìg of PGE_1. Two terrestrial arthropods, the scorpions *Bathus occitanus* and *Androctonnus australis*, produced remarkable fevers of 15 to 20°C increases in temperature after treatment with physiological doses of PGE_1. They regulated their body temperatures by placing themselves in selected positions along temperature gradients in a sand box.

The observed increases in body temperatures after treatments with individual doses of PGE_1 are not conclusive on the roles of PGs in mediating fevers in invertebrates. A number of questions need answers to help place this work on firmer ground. Do the appropriate PGs naturally occur in these species? Is the fever response specific to PGE_1? Does PG biosynthesis increase after infection? These points remain open, but they may assume considerable interest if induction of febrile responses is a major component of invertebrate immunity.

A wide range of animals, including a large number of insect species, regulate their body temperatures by behavioral means, such as basking in the sun. Many insect species utilize behavioral and physiological means to maintain constant body temperatures in parts of their bodies while they are exposed to a wide range of ambient temperatures.

Queen bumblebees provide an example of physiologically regulating the temperature of parts of their bod. When flying at higher temperatures, the queens regulate the patterns of hemolymph circulation to favour transfer of heat from the thorax, where it is produced by the prodigious flight muscle energy metabolism, to the abdomen, from whence it can be radiated away from the body.

During flight at low temperatures, hemolymph circulation is regulated such that heat is not transferred out of the thorax. Beyond regulating thoracic temperatures during flight activity, during the early days of the colony cycle, queens enhance developmental rates by incubating their brood clump.

Heat is generated in the flight muscles, which are uncoupled from wing movements. Again, hemolymph circulation is regulated to favour movement of heat from the thorax to the abdomen which is adpressed to the brood clump during incubation to facilitate transfer of a high proportion of the heat to the developing brood. Albeit with some differences in specific pattern, many of the insects that physiologically regulate their thoracic temperatures do so by altering hemolymph circulation.

Some insects reduce their body temperatures by evaporative cooling. Cicadas are unique in their ability to regulate their body temperatures by transporting water across their cuticles, which cools the animal by evaporation from the body surface. This sweating has been analyzed in some detail in cicadas of genera *Diceroprocta* and *Tibicen*. These cicadas allow their body temperatures to rise to a particular level, at which point they actively transport water via specialized tissues to their body surfaces. The temperatures at which water transport begins are called the temperature set points.

Toolson, Ashby, Howard and Stanley-Samuelson (in preparation) suggest that eicosanoids play a number of roles in the physiology of sweating in cicadas. The cicadas were treated with inhibitors of PG biosynthesis by intrahemocoelic injection, and the effects on the temperature set point and water loss rates determined. Treating cicadas with the cyclooxygenase inhibitors aspirin and paeonol resulted in significantly higher water loss rates and lower thoracic temperatures, compared to control animals.

Treatment with 1-series PGs produced small, but significant increases in temperature set points. Injecting the precursor to 1-series PGs, 20:3n-6, increased water loss rates, but did not impact body temperature. These results indicate that PGs are mediators of thermoregulatory functions in normothermic cicadas.

Body temperature is the outcome of a complex network of events. In cicadas, body temperature is thought to minimally involve sensing and integrating body temperature information, control of thermoregulatory behaviors, regulation of circulation patterns, and control of transport events within sweat glands. Future work is aimed at understanding possible roles of eicosanoids within these systems. Because eicosanoids may be involved in a number of these areas, a direct relationship between eicosanoid biosynthesis and organismal-level temperature regulation is not expected.

As in many other insect species, the occurrence and metabolism of eicosanoid-precursor fatty acids has only recently been considered in cicadas. Gas chromatography-mass spectrometric analysis confirmed that the PLs of selected tissues, including thoracic muscles, abdominal dorsal glands, Malpighian tubules, guts and ovaries include small proportions of 20 : 3n-6, 20 : 4n-6 and 20 : 5n-3. T

hese small proportions are consistent with findings from other terrestrial insects. Male and female adult cicadas were able to convert exogenous radioactive arachidonic acid into three PGs, PGD_2, PGE_2 and $PGF_{2á}$, as well as other unidentified products. The radioactive products were identified by a single thin-layer chromatography step, so the PGs should be regarded as tentatively identified. Nonetheless, we have sufficient evidence to postulate that eicosanoids are important in the thermobiology of these cicadas.

MODULATION OF LIPID MOBILIZATION

Several hormones, including glucagon and epinephrine, stimulate mobilization of lipids in mammalian adipose tissue. The effects of these hormones are modulated by locally produced PGs: PGE_2 reduces the amount of lipid that is mobilized in response to epinephrine. The PG effect is thought to be exerted through action on adenylate cyclase.

Two studies indicate that PGs similarly modulate lipid mobilization in insects. Several reviews detail the roles of adipokinetic hormone in maintaining homeostasis of hemolymph lipids. One action of adipokinetic hormone is the release of diacylglycerol from fat body cells. In studies with *in vitro* locust fat body preparations, Wagemans, van der Horst and Stanley-Samuelson examined the interaction of PGE_2 with adipokinetic hormone.

Fat body preparations were treated with 10 pmol of adipokinetic hormone and increasing doses of PGE_2, then the amount of lipid released into the medium was assayed. In these experiments, addition of adipokinetic hormone resulted in the release of 1004 ìg of lipid. When added to the fat body preparations in doses from 5×10^{-5} to 5×10^{-3}, PGE_2 attenuated the hormone effects. These results support the hypothesis that PGs modulate lipid mobilization in insects.

Yamaja Setty and Ramaiah presented results indicating that PGE $_1$ decreased lipid mobilization in male pupae of the silkworm *B. mori*. In their experiments, male pupae were treated with saline, PGE_1 or with PGE isolated from *B. mori* male reproductive organs. Three parameters were assayed: esterase activity, lipase activity and hemolymph lipid titre. Compared with controls, pupae that were treated with PGE_1 expressed significantly reduced fat body esterase and lipase activities, and they had significantly lower levels of hemolymph lipid. These experiments were not designed to examine the possible relationships between adipokinetic hormone and PGs, but they do suggest that PGs are one of the elements that regulate lipid metabolism in insects. PGs may regulate lipid metabolism in two ways. One is the direct down-regulation of lipid mobilization as just described, and the other is through interaction with adipokinetic hormone.

As in other potentially important physiological roles of eicosanoids in invertebrates, a number of questions need attention. For one example, do insect fat bodies produce PGs? Work on another line indicates that PGs and possibly other eicosanoids are biosynthesized by microsomes prepared from fat bodies of the cockroach *P. americana*.

Given that fat body is generally competent to biosynthesize eicosanoids, and the PGE_1 impacts on lipid metabolism, further investigation of the roles of eicosanoids can be expected to yield important insights into regulation of lipid metabolism in insects.

EICOSANOIDS IN THE BIOLOGY OF OTHER INVERTEBRATES

The foregoing sections implicate eicosanoids in one or another aspect of immunology, ion transport, reproduction, thermobiology and lipid biology in certain insect species. Some eicosanoid actions do not occur in all or even many insect species. Releasing egg-laying behaviour appears to be a specialized action, so far seen in a few cricket species and in the silk moth.

Conversely, regulation of ion transport and mediation of cellular immune responses may be fundamental eicosanoid roles in many animal phyla, vertebrate and invertebrate. We can gain a greater appreciation of the fundamental roles of eicosanoids by drawing on information from invertebrates other than insects and on non-mammalian vertebrates.

EICOSANOIDS IN INVERTEBRATE IMMUNITY

In addition to their role in mammals, there is increasing recognition of the roles of eicosanoids in the immune responses of other organisms. Pettitt *et al.* reported that macrophages from rainbow trout produce lipoxins as major products, as well as small amounts of mono-hydroxy fatty acids and two leukotrienes. These products may relate to cellular migration in wound and infection responses, because Hunt and Rowley showed that fish leucocytes express enhanced migration in response to leukotriene B4.

These and other findings indicate that eicosanoids are important in immunity of all vertebrates. More recently, Hampson*et al.* reported that blood cells of the crab *Carcinus maenas* synthesized lipoxygenase, cyclooxygenase and cytochrome P-450 monooxygenase products. While we do not yet know that these products are directly involved in crab immunity, it is reasonable to suppose they are so involved. Taken with our work on insect immunity, these results suggest that eicosanoids may be involved in the immune responses of many invertebrate animals.

EICOSANOIDS IN INVERTEBRATE REPRODUCTION

Eicosanoids modulate reproductive physiology and behaviour of invertebrates from several phyla. The hatching factor of the barnacle *Balanus balanoides* was identified as the novel eicosanoid 10, 11, 12trihydroxy-5, 8, 14, 17-eicosatetraenoic acid. Similarly, 8-HETE is the hatching factor of another barnacle, *Eliminius modestus*. Aside from Arthropoda, eicosanoids in many aspects of molluscan reproduction. Among bivalves, Osada*et al.* implicated PGs in spawning in the scallop *Patinopecten yessoensis*.

PGs are also involved in gastropod reproduction. PGE_2 increased egg production in the snail *Helisoma duryi*, and PGs accelerate the onset of egg laying in the pond snail *Lymnaea stagnalis*. Among Echinodermata, the eicosanoid (8R)HETE, induces oocyte maturation in some, but not all, starfish. In their detailed studies of the sea urchin model of fertilization, Schuel*et al.* indicated that eicosanoid biosynthesis is one of several redundant mechanisms involved in preventing polyspermic fertilizations. A prominent (11R) and (12R) lipoxygenase activity occurs in eggs of the sea urchin.

Taken together with the earlier discussion of eicosanoids in nonmammalian vertebrates, these examples indicate that eicosanoids are major regulators of reproductive biology in vertebrate and invertebrate animals. These regulators act at organismal, tissue, cellular and subcellular levels. Expression of some eicosanoid functions, such as release of egg-laying behaviour and ovulation pheromone signaling, is likely to be restricted to a relatively small number of animal species.

Other roles, such as prevention of polyspermic fertilizations, appear to be fundamental actions that may occur throughout Animalia. When viewed on the broader landscape of animal reproduction, it is not surprising to see that eicosanoids occur in the reproductive tissues of insects. Exploring the biological meaning of these compounds will undoubtedly reveal hitherto unseen levels of regulation in the physiology of insect reproduction.

EICOSANOIDS IN INVERTEBRATE NEUROPHYSIOLOGY

PGs have been detected in extracts of brains from crickets and from ticks, as well as in extracts of heads and thoraces from seven other common laboratory insect species. Stanley-Samuelson and Loher put forth the idea that in releasing egg-laying behaviour in the Australian field cricket, PGE_2 exhibits a hormonal mode of action. They suggested that egg-laying behaviour results from the interaction of PG with still unidentified elements of the cricket nervous system, although the nature of the interaction is not clear.

Nanda and Ghosal observed that injecting (site unspecified) females of the cockroach P. americana with 1.25 ìg $PGF_{2á}$ /animal resulted in reduced amounts of neurosecretory material in the neurosecretory cells of the pars intercerebralis as observed under a light-microscope.

They suggested that PGs are involved in the physiology of secretion in these cells. Aside from these ideas, there is no direct evidence that PGs are involved in insect neurobiology. However, findings from the sea hare *Aplysia* indicate that eicosanoids are fundamental elements in synaptic physiology.

The peptide FMRFamide (a tetrapeptide denoted by single letter amino acid code) is active in sensory neurons of *Aplysia*. This molecule produces effects opposite to the effects of 5-HT, that is, it hyperpolarizes the nerve cell membrane, decreases action potential duration, and increases the outward potassium current, among other actions. Piomelli *et al.* showed that arachidonic acid and the lipoxygenase metabolite 12-HPETE mimic the actions of FMRFamide.

They demonstrated that neural tissue biosynthesizes lipoxygenase products, and that inhibition of phospholipase activity suppresses the FWRFamide effects. The authors concluded that the release and metabolism of arachidonic acid mediate the inhibitory response to FMRFamide in *Aplysia* sensory neurons. Further characterization of the 12-lipoxygenase pathway showed that the 12-HPETE is metabolized to 12-keto-5, 8, 10, 14 eicosatetraenoic acid (12-KETE). The authors suggested that 12-HPETE and its metabolite 12-KETE participate in synaptic actions in the *Aplysia* nervous system.

To the extent that *Aplysia* models the details of neurophysiology in animals, these results suggest that eicosanoids are fundamentally involved in synaptic transmission. From the point of view of our efforts to gain a thorough appreciation of insect physiology, the findings from this molluscan model argues that, as discussed elsewhere, eicosanoids appear to represent a hitherto unseen tier of regulatory elements.

BLOOD FLUKES: SKIN PENETRATION BY CERCARIAL LARVAE

The roles of eicosanoids in blood flukes were reviewed elsewhere, from which these abbreviated remarks are taken. Eggs of the blood fluke *Schistosoma mansoni* (Platyhelminthes, class Trematoda) leave their mammalian hosts along with the feces. If they are deposited in water, the eggs hatch and continue larval development in snails. After leaving their intermediate hosts, freeswimming cercarial larvae infect mammalian hosts by ingestion with drinking water or by burrowing through the skin.

Among skin lipids, free PUFAs are most efficacious in stimulating cercarial penetration. Cercarial penetration is the outcome of two behaviors, cessation of swimming and initiation of penetration. Both behaviors are mediated by eicosanoids, rather than by the PUFAs themselves.

In studies of cercarial penetration into membranes prepared from EFA-deficient and EFA-replete rats, there was about threefold less penetration into membranes from the EFA deficient rats. Interperitoneal injections of the cyclooxygenase inhibitor ibuprofin led to time-dependent accumulation of the drug in the skin of EFA-replete rats. Cercarial penetration into membranes prepared from these rats was reduced in proportion to the amounts of ibuprofen that accumulated in the skin. Under conditions that reduce the levels of PUFAs and reduce eicosanoid biosynthesis, cercarial penetration is reduced.

Cercariae are competent to synthesize eicosanoids. When cercariae were incubated with radioactive linoleic acid, radioactivity was recovered in HPLC fractions that corresponded with PGE_2, PGD_2, LTC_4, LTB_4 and 5-HETE. Fusco*et al.*, concluded that formation of eicosanoids is an essential step in penetration of human skin by cercariae of *S. mansoni*.

Thus, eicosanoids appear to be involved in three aspects of the relationship between blood flukes and their mammalian hosts, specifically cessation of swimming, initiation of penetration behaviour and the possible modulation of the host immune system by worm-derived PGs.

These and other examples of eicosanoid actions in invertebrates are presented here to place the physiology of insects onto the broader landscape of invertebrate physiology. To the extent that insects fit into this context, the information from invertebrates can provide insights into insect physiology.

TESTING THE MAMMALIAN EICOSANOID MODEL IN INSECTS

Elements of the mammalian model of eicosanoid biology include the biosynthesis and occurrence of eicosanoid-precursor PUFAs, release of PUFAs from cellular phospholipids, eicosanoid biosynthesis and mode of action, and catabolism of eicosanoids. Because the vast majority of our information on eicosanoids comes from research on mammalian systems, there is a tendency to channel our thinking along the guidelines that are established from these systems.

How well do the eicosanoid systems in invertebrates fit the mammalian model? The significance of this question lies in how we form our research questions and our experimental approaches. If our expectations are that the biochemistry and actions of eicosanoids will fit the mammalian model in an exact way, our experimental designs could be fatally flawed.

To help appreciate some of the assumptions that we form by overreliance on the mammalian model, comparisons drawn between the mammalian model and invertebrate systems in the following paragraphs.

EICOSANOID PRECURSOR POLYUNSATURATED FATTY ACIDS

Eicosanoid biosynthesis depends upon the availability of the appropriate precursor PUFAs, especially arachidonic acid. Although the point is solidly established in mammals, we have only recently come to recognize the occurrence of these fatty acids in insects.

Terrestrial herbivorous insect species differ from mammals on the point that arachidonic acid and other C20 PUFAs occur in extremely low quantities (often 0.1% of PL fatty acids) in many of the terrestrial insects. This is not so for all insects, because aquatic and some predatory insect species feature quite high proportions (7 - 25% of PL fatty acids) of C20 PUFAs. Uscian and Stanley-Samuelson (unpublished observations) considered the possibility that terrestrial arthropods other than insects might also be quite low in these fatty acids.

We analyzed the phospholipid fatty acids prepared from a number of spiders, scorpions, and centipedes and found that all of these terrestrial arthropods contained considerably higher proportions of C20 PUFAs than did terrestrial insects. We surmised that the very low proportions of these fatty acids are unique to most terrestrial insects, however, there is enough information to support the view that C20 PUFAs are generally present in insect lipids.

Phospholipase A_2

Eicosanoid biosynthesis is thought to begin with hydrolysis of the appropriate C20 PUFA from cellular phospholipids. C20 PUFAs are preferentially associated with the *sn*-2 position of phospholipids, and PLA_2 activity is thought to be a regulating step in eicosanoid biosynthesis. We have very limited information on PLA_2 activity in invertebrates. Ragab*et al.* described a PLA_2 activity in reproductive tissues of the firebrat *Thermobia domestica* and indicated that eicosanoid synthesis was correlated with PLA_2 activity in these tissues. These findings show that firebrat reproductive tissues are similar to the mammalian model on the role of PLA_2 in eicosanoid biosynthesis.

PLA_2 activity in the salivary glands of the lone star tick has been investigated by Sauer and his colleagues. The work here because this is the most detailed research on PLA_2 activity in any invertebrate. It appears that the tick salivary gland is also similar to mammals on the point of the role of PLA_2 activity in eicosanoid biosynthesis.

We have been investigating PLA_2 activity in the tobacco hornworm fat body (Uscian and Stanley-Samuelson, in preparation). Fat body homogenates are incubated with PC labeled with [1-^{14}C]arachidonic acid in the *sn*-2 position. After selected incubation periods, total lipids are extracted and the free fatty acids are separated from phospholipids by TLC. PLA_2 activity is calculated from the amount of radioactivity in the free fatty acid fraction. These experiments show that the fat body contains a cellular PLA_2 activity that can hydrolyze arachidonic acid from the *an*-2 position of PC.

A composite of data from four lines of experiments which indicate that PLA_2 activity is dependent on homogenate protein concentration, time, temperature and pH. PLA_2 activity increased linearly with increasing homogenate protein concentrations up to 0.8 mg protein. This pattern is similar to results from homogenates of rat mesangial cells, but quite different from results of similar experiments on homogenates prepared from a macrophage cell line.

The fat body $PILA_2$ activity is different from most PLA_2 s because it appears to be either calcium-independent or else requires very low levels of free calcium. Many forms of PLA_2, particularly the 14 kDa forms, require millimolar calcium concentrations to reach their maximal rates of hydrolysis. Calcium-independent PLA_2 s occur in guinea pig intestinal brush border membranes, among other sources. There are also PLA_2 s from a number of mammalian tissues that are activated by micromolar calcium concentrations (10-7-10-6M). These are the soluble, high molecular weight (60 - 110 kDa) forms of PLA_2. The fat body activity we observed may be similar to either of these mammalian PLA_2s. It would appear from the information on the firebrat, the lone star tick and on the tobacco hornworm that invertebrates are similar to the mammalian model on the point of having PLA_2 activities that can hydrolyze arachidonic acid from the *an*-2 position of cellular phospholipids. However, it remains to be rigorously shown that eicosanoid biosynthesis depends on the PLA_2 activity we recorded.

EICOSANOID BIOSYNTHESIS IN TOBACCO HORNWORMS

The idea that eicosanoids mediate the cellular immune responses of an invertebrate animal opens a novel and very broad field for investigation. One of the more pressing questions relates to the identities of the active eicosanoids: Which of many possible eicosanoids are biosynthesized by immune tissues? We are currently addressing this question in fat body and hemocytes from tobacco hornworms. A detailed report of the information presented here in outline form is forthcoming.

Microsomal-rich fractions were incubated with radioactive arachidonic acid, the reactions stopped, and the eicosanoid products extracted with acidified ethyl acetate. The reaction products were separated by 1-dimentional and 2-dimentional TLC, and the radioactivity in each fraction was assessed by liquid scintillation counting. The experiments were preceded by a 3 min pre-incubation with all reaction components except the protein source in order to bring all components to the incubation temperature before the experiment begins.

This is an important step because, compared with other studies of insect PG biosynthesis, we use far shorter incubation periods. After spotting the samples, the TLC plates were developed in solvent system A9 and the fractions visualized by exposure to iodine vapors.

CONTROL EXPERIMENTS

PUFAs are labile to spontaneous reactions with oxygen, which often creates reaction artifacts. These possibilities were accounted for by conducting "zero-time" reactions in which the substrate is pre-incubated in parallel with our biosynthesis experiments.

At time zero, the preincubations are stopped and the extraction and analysis of the products followed the same procedures just described for the biosynthesis experiments. Values from these control experiments were used to correct the corresponding values in the PG biosynthesis experiments.

Microsomal-rich preparations from the fat bodies isolated from fifthinstar tobacco hornworm larvae synthesize substantial amounts of four prostaglandins, $PGF_{2á}$, PGE_2, PGD_2 and PGA_2. Under most conditions, PGA_2 was the major product, although we will show that the incubation conditions alter the overall pattern of PG biosynthesis. The effects of altering our standard in vitro conditions are presented in the following paragraphs.

INFLUENCE OF MICROSOMAL PROTEIN LEVEL ON PROSTAGLANDIN BIOSYNTHESIS

The relationship between quantities of fat body microsomal protein and PG biosynthesis. These data indicate that the optimal amount of microsomal protein for prostaglandin biosynthesis was about 0.75 to 1.0 mg. Based on these findings, we used 1.0 mg of microsomal protein in subsequent experiments. These results were obtained from 10 min incubation periods, which we next show is not the optimal incubation period.

SHORT INCUBATIONS ARE OPTIMAL FOR PROSTAGLANDIN BIOSYNTHESIS

The profiles of PG biosynthesis at 10 incubation intervals from 0.1 to 10.0 minutes. The largest total PG biosynthesis (3,400 pmol/mg protein) appeared in 1.0 minute incubations. Total PG production was diminished in longer incubation periods, from about 2,100 pmol/mg protein at 2 min to about 400 pmol/mg protein at 10 min. Also, total PG biosynthesis was reduced in very short incubation periods, those less than 1.0 min.

Although PGA_2 was the major product in most incubation periods, the duration of the incubation appeared to alter the overall profiles of PG biosynthesis. At 0.1 min, the shortest incubation, $PGF_{2á}$, was the only detectable product. At 0.25 min, all four PGs were formed with PGF_2 again the major product (1,000 pmol/mg protein), while PGA_2 biosynthesis was just detectable (ca. 20 pmol/mg protein).

However, at all incubation periods from 0.5 to 10 min, PGA_2 was the major product, by 3 to 10-fold. Substantial amounts of PGD_2 were produced in all the incubation periods beyond 0.1 min; PGD_2 synthesis varied with incubation protein at 35°C and to < 150 pmol/mg protein at 40°C. Negligable biosynthesis occurred at 45°C.

The temperature effects recall our theme of the effects of incubation conditions on biosynthesis of individual components. At 25°C, PGD_2

was the major product (> 400 pmol/mg protein compared to about 180 pmol/mg protein for PGA_2), while at 30°C, biosynthesis of PGA_2 was about an order of magnitude greater than of PGD_2 (> 1700 compared to about 160 pmol/mg protein).

Although it was produced in lower abundance at the lower temperatures, $PGF_{2\acute{a}}$ was the major PG, while PGA_2 synthesis was reduced at 35°C. $PGF_{2\acute{a}}$ biosynthesis appears to be quite temperature sensitive, because this product was not synthesized at 40 and 45°C.

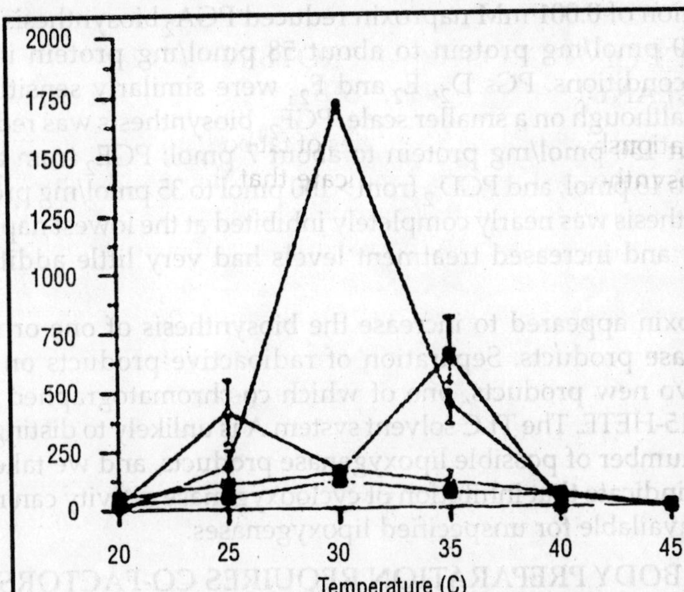

Fig. Eicosanoid Biosynthesis by Tobacco Hornworm Larvae Fat Body Preparations

PGE_2 IS NOT RAPIDLY CONVERTED TO PGA_2 UNDER OUR STANDARD CONDITIONS

PGA_2 biosynthesis is thought to proceed from PGH_2 through PGE_2, which can be converted to PGA_2 and PGB_2; PGA_2 is not directly formed from PGH_2. The relatively high levels of PGA_2 biosynthesis by fat body preparations could be an artifact of spontaneous rearrangements of PGE_2 during our analytic steps, a product of nearly quantitative PGE_2 metabolism, or a product of an unknown biosynthetic route that has not yet come to light.

We investigated these possibilities by incubating fat body preparations with radioactive PGE_2. About 90% of the radioactivity was recovered in a fraction that co-chromatographed with authentic

PGE_2, while 3%, 4% and 2.6% was recovered as $PGF_{2á}$, PGD_2 and PGA_2, respectively. These data indicate that PGE_2 is not quantitatively converted to PGA_2, and that the analytical procedures did not produce PGA_2 as an artifact. Our working hypothesis is that the hornworm fat body produces PGA_2 by way of an unidentified intermediate.

FAT BODY PREPARATION IS SENSITIVE TO INDOMETHACIN

Addition of 0.001 mM naproxin reduced PGA_2 biosynthesis from over 1,700 pmol/mg protein to about 58 pmol/mg protein in our standard conditions. PGs D_2, E_2 and $F_{2á}$ were similarly sensitive to naproxin, although on a smaller scale. $PGF_{2á}$ biosynthesis was reduced from about 150 pmol/mg protein to about 7 pmol; PGE_2 from about 115 pmol to 13 pmol, and PGD_2 from > 160 pmol to 35 pmol/mg protein. PG biosynthesis was nearly completely inhibited at the lowest naproxin treatment, and increased treatment levels had very little additional influence.

Naproxin appeared to increase the biosynthesis of one or more lipoxygenase products. Separation of radioactive products on TLC yielded two new products, one of which co-chromatographed with authentic 15-HETE. The TLC solvent system A is unlikely to distinguish among a number of possible lipoxygenase products, and we take this finding to indicate that inhibition of cyclooxygenase activity can make substrate available for unspecified lipoxygenases.

THE FAT BODY PREPARATION REQUIRES CO-FACTORS USED IN MAMMALIAN PREPARATIONS

PG biosynthesis has been investigated in a number of insect and other invertebrate systems. The general procedures usually include incubating a tissue homogenate with radioactive substrate, then monitoring eicosanoid biosynthesis by a form of radio-chromatography.

Based on experience with mammalian preparations, most experiments with insects were performed in a buffer amended with a cocktail of co-factors, usually a mixture of reduced glutathione and hematin. Hemoglobin or myoglobin is often used in place of hematin. The roles of these co-factors relates to the mechanism of cyclooxygenase activity. In mammals, PGH synthase is a combination of two activities, a cyclooxygenase and a hydroperoxidase, in a single holoenzyme molecule. Cyclooxygenase first converts a PUFA to PGG and then reduced glutathione is oxidized by the attendant hydroper-oxidase in the process of reducing PGG to PGH.

The cyclooxygenase and hydroperoxidase activities require heme, specifically iron protoporphyrin IX. The globins, myoglobin and hemoglobin, also stimulate both activities of PGH synthase *in vitro* although the stimulation is due to release of heme from the globins.

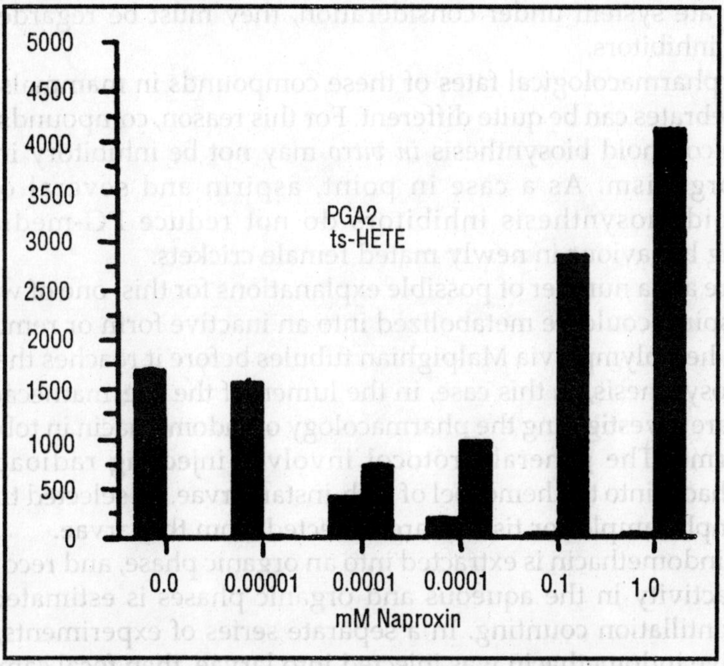

Fig. Eicosanoid Biosynthesis by Tobacco Hornworm Larvae Fat Body Preparations

If the PG biosynthetic enzymes in insects are similar to their counterparts in mammals, we would expect that the insect preparations would similarly require heme and reduced glutatione. These co-factors are included in incubations with insect preparations, however, it is not known that they are required for eicosanoid biosynthesis.

We used the fat body preparation to directly test this idea, and found that PG biosynthesis was reduced to zero or nearly zero in the absence of the co-factor cocktail (Stanley-Samuelson and Ogg, in preparation). We do not yet know if the heme binds to insect and mammalian cyclooxygenases in the same way. Nevertheless, we conclude that the insect PGH synthase is similar to its mammalian counterpart on the requirement for co-factors.

Pharmacology of Indomethacin in Tobacco Hornworms

Eicosanoid-biosynthesis inhibitors have been used to probe roles of eicosanoids in many studies of invertebrates. However, these inhibitors are functionally defined on the basis of appropriate investigations of mammalian systems, largely motivated by clinical requirements. The caveat to offer is that until the inhibitors have been characterized as eicosanoid biosynthesis inhibitors in the particular invertebrate system under consideration, they must be regarded as putative inhibitors.

The pharmacological fates of these compounds in mammals and in invertebrates can be quite different. For this reason, compounds that inhibit eicosanoid biosynthesis *in vitro* may not be inhibitory in the whole organism. As a case in point, aspirin and several other eicosanoid-biosynthesis inhibitors do not reduce PG-mediated egglaying behaviour in newly mated female crickets.

There are a number of possible explanations for this, one of which is that aspirin could be metabolized into an inactive form or removed from the hemolymph via Malpighian tubules before it reaches the site of PG biosynthesis, in this case, in the lumen of the spermathecae.

We are investigating the pharmacology of indomethacin in tobacco hornworms. The general protocol involves injecting radioactive indomethacin into the hemocoel of fifth-instar larvae. At selected times, hemolymph samples or tissues are collected from the larvae.

The indomethacin is extracted into an organic phase, and recovery of radioactivity in the aqueous and organic phases is estimated by liquid scintillation counting. In a separate series of experiments, the radioactive indomethacin was injected into larvae, then fecal samples were collected every 4 hours for the next 48 hours. Estimating recovered radioactivity followed the procedure just described.

There are three salient findings on the distribution of radioactive indomethacin in the hornworms. First, over 99% of the injected indomethacin was cleared from hemolymph circulation within 3 minutes. Second, radioactive indomethacin was recovered in all tissues analyzed; gut wall, gut contents, Malpighian tubules, fat body, ventral nerve cord, hemolymph, silk gland and integument. Fat body, the largest tissue, accumulated nearly half of the recovered radioactivity. However, when bioaccumulation was normalized to tissue weight, the silk glands accumulated most of the recovered radioactivity. The third finding is that the radioactivity recovered from these tissues co-chromatographed with authentic indomethacin on TLC.

Mammals convert indomethacin into three compounds, des-chloroindomethacin, methchloroindomethacin, benzoylindomethacin, all more polar products that can be excreted in urine. Tobacco hornworms also produced a polar product, however, the product is not as polar as the products of mammalian metabolism. Moreover, the polar product was recovered from the frass, not from any of the internal tissues. This finding opens the possibility that indomethacin is subject to microbial metabolism after it is secreted into the gut contents.

These findings are germane to investigations of PG actions in insects. The PG biosynthesis in spermatophore preparations from the cricket *A. domesticus* differed from the mammalian preparations on two points. First, the cricket PG biosynthetic activity was associated with the 12,100 *g* pellet, rather than the microsomal fraction. Second, the cricket preparation was insensitive to indomethacin.

Destephano*et al.* showed that 0.001 M indomethacin nearly completely inhibited PG biosynthesis in bovine seminal vesicle preparations, but had no effect on the 12,100 *g* pellet from cricket reproductive tracts. This is an important point, because many eicosanoid-biosynthesis inhibitors are used as probes to examine the possible roles of eicosanoids in invertebrate systems on the assumption that the inhibitors that have been thoroughly characterized in mammals will have similar functions in invertebrates. The situation is probably otherwise.

MODES OF EICOSANOID ACTIONS

Eicosanoids are thought to exert their actions through interaction with specific receptors located on the outer surface of cell membranes. Due to their potential applications in clinical settings, mammalian eicosanoid receptors have received considerable attention. Under the general model, eicosanoid receptors are coupled to intracellular second messenger effectors through various G proteins. The second messenger systems are not the same for all eicosanoids.

E-series PGs generally influence intracellular cAMP concentrations, in some cases inhibiting and in other cases stimulating adenylate cyclase. The $PGF_{2á}$ receptors influence phospholipase C, but seem to have no effect on cAMP. Negishi *et al.* suggested that the adrenal medulla PGE receptors are coupled to two signal transduction systems: an adenylate cyclase via G_i and to production of inositol phosphates via a pertussis toxin-insensitive G protein. PGE receptor subtypes exist in many mammalian tissues, which would account for two different signal transduction systems for a single ligand.

The foregoing remarks are meant to outline one element of the mammalian model of eicosanoid biology. Smith provides a good overview of the modes of eicosanoid actions; Colman*et al,* provide detailed treatments of this subject.There is little evidence on eicosanoid receptors in invertebrate animals.

Freas and Grollman demonstrated specific PGA_2 receptor sites in homogenates of several tissues from the marine bivalve *Modiolus demissus*. These studies were conducted in the context of regulating ion flux, which is treated in detail by Stanley-Samuelson. In gills, the PGA_2 receptors were pH dependent and reversible. This remains the only direct examination of a PG receptor in any invertebrate. Stanley-Samuelson*et al.* developed circumstantial evidence on PGE receptors in females of the Australian field cricket *T. commodus*.

Studies of the relationship between eicosanoid structures and release of oviposition behaviour showed that the effects on egg-laying are specific to E-series PGs. This specificity indicates that the egg-laying effect of PGE is mediated through receptors that can distinguish the E-series PGs from many other closely related structures. Morton and Giunta suggested that eclosion hormone stimulates a membrane-bound form of guanylate cyclase in an indirect manner.

They showed that inhibition of the release of arachidonic acid from phospholipids and inhibition of arachidonic acid metabolism abolished the effects of eclosion hormone on guanylate cyclase. These findings indicate that eicosanoids may be part of the hormone signal transduction pathway. If this is so, then transmembrane signaling may be another mode of eicosanoid action in insects.

Index

D

E

F

G